THE DEVELOPMENT OF DESIGN

THE DEVELOPMENT OF DESIGN

GORDON L. GLEGG

CONSULTING ENGINEER AND FORMERLY LECTURER IN ENGINEERING
UNIVERSITY OF CAMBRIDGE

CAMBRIDGE UNIVERSITY PRESS

CAMBRIDGE

LONDON · NEW YORK · NEW ROCHELLE

MELBOURNE · SYDNEY

CAMBRIDGE UNIVERSITY PRESS
Cambridge, New York, Melbourne, Madrid, Cape Town, Singapore, São Paulo, Delhi

Cambridge University Press
The Edinburgh Building, Cambridge CB2 8RU, UK

Published in the United States of America by Cambridge University Press, New York

www.cambridge.org
Information on this title: www.cambridge.org/9780521113175

First published 1981
This digitally printed version 2009

A catalogue record for this publication is available from the British Library

ISBN 978-0-521-23794-9 hardback
ISBN 978-0-521-11317-5 paperback

CONTENTS

INTRODUCTION

Most people regard scientists as explorers; and this is always true: or very nearly always true.

Imagine a handful of people shipwrecked on a strange island and setting out to explore it. One of them cuts a solitary path through the jungle, going on and on until he is exhausted or lost or both. He eventually returns to his companions, and they listen to him with goggling eyes as he describes what he saw; what he fell into, and what bit him. After a rest he demands more supplies and sets off again to explore the unknown. Many of his companions will be doing the same, each choosing his own direction and pursuing his pioneering path.

The shipwrecked explorers of such an unmapped island and the scientific explorers of our strange and unpredictable universe have some characteristics in common. Neither group ever reaches a brick wall where no further advance in knowledge is possible. A scientist may exhaust himself; he frequently exhausts his colleagues, always exhausts his money, but never exhausts his subject. Sighing for new worlds to conquer is not an occupational hazard of his profession. The second common characteristic of both types of explorers is that their narrow paths rarely cross. Spoke-like, their trails into the unknown leave the little hub of common knowledge far behind and their fellow explorers further and further out of touch.

Scientists in differing subjects do not rub each other up the wrong way for they are not in rubbing distance; they may be touchy but they don't touch.

In his inaugural lecture on reconciling Physics with reality, Professor A. B. Pippard, FRS, Cavendish Professor of Physics at Cambridge said 'I think we may picture those domains where understanding exists, whether in physics, chemistry, biology, psychology, economics or any other discipline as cultivated valleys in a formidably mountainous country. We may recognise in principle that we all inhabit the same world but in practice we do well to cultivate our own valleys, with an occasional assault on the more accessible foothills, rather than to build roads in a vain attempt at colonisation.'

Now let us go back to the island and see how the castaways are progressing. The explorers are having fun; the rest are not. Those left on the sea shore desperately need shelter, food and warmth. Talking about 'vain attempts at colonisation' does not comfort them at all; colonisation is precisely what they urgently need. 'Can't someone stop exploring the unknown and make the known a little more habitable, clear an area of jungle by the sea shore and build us some comforts?' they say as, hopelessly, they watch each explorer, in turn, rabbiting off again into the undergrowth. Is the position hopeless or is there an exception to the rule; can they find an explorer who stops at home? There is one and only one; he is the engineering scientist. He not only seeks knowledge but he also applies it. His duty is to the community. His success lies in the tangible, and his satisfaction springs from creating something both new and useful. He is the exception to the rule, the odd scientist out. Others define their research by its direction, he defines his by area. He must demolish a sizeable plot of the variegated jungle of ignorance before he can start to build.

But this is only a beginning. He must then conjure up from his resources, both mental and material what he and his community need. This statement presents, in its simplest form, the purpose of this book. More specifically, the aim is to present an over-all pattern of how engineering design is developed, hoping to give confidence to those who find the subject faintly frightening, and to focus the enthusiasm of those who find it fascinating. The aim is also to give an interpretation in the industrial context of the author's more academic books: 'The Design of Design', 'The Selection of Design' and 'The Science of Design'.

If we sit back and seek to find a comprehensive view of how design progresses from the abstract to the concrete, we may well find that we can form four different images of it; none wholly complete without the others.

The first is to analyse into separate compartments subjects that may, in practice, sometimes overlap, but this over-simplification may be an essential element in clarifying the picture or writing a book. In this view there are three subjects that are normally found, in turn, in the development of an engineering design – the basic idea, its first embodiment and its contemporary embodiment.

In the realm of the basic idea lies the happy (or unhappy) hunting ground of the inventor. Is the new idea he has dreamt up merely a pipe dream; will it, like Alice's Cheshire Cat, dissolve into nothing leaving only a tantalising grin? Only research can provide a decisive answer. The first embodiment of an idea is the realm of the prototype; the first attempt at a total working entity. The contemporary embodiment of the idea is the state of the art at the moment.

These three subjects should not be, in theory, too difficult to distinguish. The first concerns a castle in the air, the second a castle on the ground, the third a castle in the market place; but in practice it is sometimes easy to muddle the first two. A primary reason for this is that a new idea has, inevitably, to be dressed up in old technology. The appropriate technology is evolved later. The possibility of confusion lies in condemning a new basic idea because of the defects in the initial components in which it is being demonstrated.

When I was at Cambridge, a fellow student told me with glee that he had just bought himself two aeroplanes. I asked why two? Was he going to try to pilot both at once? He explained that the owners of the first machine were so overjoyed that he was paying cash for it, that they threw in the second one free of charge. He took me out to look at them. They were Gipsy Moths, constructed largely by thin lengths of wire, most of which seemed only to be attached at one end. He said he would take me up in one; which one depended on which engine was most likely to start. I said I would prefer the one whose engine was least likely to stop. Overall the technology was terrifying; the one instrument, the air-speed indicator, was reputed not to work too well. There was no dash-board clock, but my friend said that presented no difficulties as all station platforms had one. If you wanted to know the time you just flew through a railway station and looked; this had the fringe benefit of enabling you to read the name of the station too; useful as one was usually lost. But perhaps the greatest fun of all was catching rabbits. You flew over the fields a few feet above the ground, hopping over the hedges. If you saw a rabbit you flew after it and tried to assassinate it by putting a wheel on it.

Disguised in debatable technology, within that little plane lay all the great inventive ideas of heavier-than-air transport. Technology has vastly improved, but even so flying through railway stations and catching rabbits is no easier with Concordes. The essential preliminary for efficient research into a subject is to decide precisely what the subject is; otherwise you may easily confuse yourself and everyone else too.

A few years ago the government allocated £5 000 000 for research into a tracked hovercraft train. When all the money had been spent the whole project was suddenly abandoned, the staff dismissed, and the bits and pieces auctioned. This caused such a row that a parliamentary enquiry was held to discover the facts, and their 100-page report made horrifying reading. Re-arranging the facts under our three subject headings, it appears that the £5 000 000 was originally allocated by those responsible for research into a first embodiment of an idea, was received by the research team, quite sincerely, as funds for investigating a basic idea, and when a government minister visited them he apparently expected to find the finished embodiment. None of the parties appeared to share a common definition, commonly understood, about almost anything. Don't follow their example, or you may follow their fate.

This picture is useful, and may well be the best basis on which to subdivide the contents of a book, but it is essentially analytical and may, as all analysis may do, lead us a step away from reality instead of towards it. As a corrective we must turn to our second image, the industrial outlook. Here the picture is quite different. The scientist who stops at home has, at first, no home to stop in. He must build one, using such technical knowledge as is available and practical. His house will both be a home and an incarnation of technology. In engineering what you see is the technology. It is the surfaced image of the design on which it is founded. All technology must have a design formation, otherwise there is nothing to be technical about. But there is a third and deeper level too; for design itself must be based on research, either past or present. You dare not depend on luck or wishful thinking, you must have established facts for the behaviour of your materials, both individually or inter-connected. The industrial picture here is therefore of a three-layered entity, a visible edifice, based on a two-layer foundation. Periodically new technologies are evolved, the existing building is bulldozed away, and an apparently completely new one takes its place. We think everything must be new, but quite often the old foundations are still left intact, or only modified in a minor way. Technology is evolving so fast that often it is out-of-date almost before it appears, but a good basic design can last a long time. You change the dress but not the girl.

The third image is that seen by the academic as he looks at the industrial hurly-burly. For him technology is industry doing its stuff. Stuff that can easily be specified and recorded in a mass of detailed drawings, looked up in handbooks, or soon found by skimming through advertisements in journals. At the design level it is quite another matter; he may well be puzzled about whether design, as such, can really be taught at all. Going lower still he may well decide that research is totally unteachable in an abstract way. There may be two reasons for thinking this – it may be too simple, or too complicated. Too simple in the sense that 'all that you have to do is try it and see'. Too complicated in the sense that all real research is breaking new ground, or pot-holing down a new hole, and unless we know what we are going to find we cannot know how best to look for it, and if we knew we wouldn't have to look anyway. Nature is too complicated, unpredictable, and surprising to be fettered to a system. For the academic therefore the surface of the subject may be beneath his dignity but the foundation over his head.

The engineer again has his own view. His image, our fourth, is fundamentally different from chess board analysis, or piled up operating areas, for he is concerned with movement. He sees a rapidly flowing stream of progress into which he dives, searching to find a pearl of inventiveness and then to surface safely with it, through intermingling currents of research design and technology; from the abstract darkness into the vividness of the fresh air world; exciting, exhausting, sometimes rewarding, often frustrating but never static; a career for those who hate being bored. But for writing a book the chopping up into subject sequences is a less volatile and an easier indexed system; but the more realistic image of the engineer should always be at the back of our minds.

This book concentrates largely on the first two subjects; the basic idea and its first embodiment. The production embodiment, the final chapter in the saga that begins as an over-simplified picture in the imagination of the inventor and climaxes in a concrete contribution to the community, is much less easy to consider in general terms. It is so dependent on the personnel, profits, policies and politics, and sometimes the peculiarities, of particular firms.

However, I hope that the following pages may assist anyone who has a design to develop, and those who are there to help him, in an industrial firm or a University Department.

I

The basic idea—direct tests

Whether an inventor is employed by a company, or hopes to be, his urgent and itching need is to know if his new idea is a valid one. And the classic, obvious, and apparently simple way to find out is by some form of direct test. 'Make one and see if it works', 'Seeing is believing, it leaves no room for argument' and so on. All this seems so obvious as to be hardly worth saying. What is not so obvious is that such statements are often expensive self-deceptions. You will be wise not to believe what you see, or what you think you see. The underlying reason for this is that the direct test method tells you what works but not why it works. It tells you something in particular but nothing in general. It has behaviour but no self-evident principles. It acts but does not talk. If you think it has a voice you are only listening to your own echo.

I was once asked to investigate a situation where a new invention behaved in a thoroughly mysterious manner. A small portable machine (whose exact purpose and design I am tactfully leaving unidentifiable) had been invented on a completely new principle. A full-size working model had been made and extensively tested both for performance and reliability. Based on these results a large company signed a favourable royalty agreement with the inventor and production plans were laid on. The machine in its production form was to be identical to the tested one with the exception of a minor dimensional adjustment felt to be essential by the sales side of the organisation. The first production machine was completed and, as a matter of routine, tested. It would not work at all. I don't mean that it just worked badly; it literally failed completely to do what was required and expected. Everyone made suggestions, none of which helped at all.

I found the problem difficult, as the working principle of the idea had basically indeterminate elements in it, but I finally established that the original model only worked because of some remarkable dimensional coincidences. It was entirely by chance that the inventor, choosing his dimensions in an almost haphazard way, had hit on this unique combination. The essential, and on the face of it unimportant, alteration of one dimension in the production model had upset the balance of everything. Quite literally so, for the contract had to be cancelled and the waiting production components scrapped.

By all means say seeing is believing but don't, in practice, necessarily believe what you see. In direct tests you see what but never why. It is all practice and no theory. And I have known other instances too. For instance a principle that only worked because the unwitting flexibility in a component allowed it to do so.

And, even more important perhaps, is the reverse of this phenomena. Valid new inventions, of great potential value to the community, have been discarded because direct tests showed that they would not work. Many new inventions are not new at all; they are re-inventions. Years before the identical idea had been buried, and 'it won't work' written on the gravestone; a good principle unjustly condemned because of some technological coincidences, failures, or inaccuracy.

The first axial flow jet engine was abandoned and the principle neglected for seven years owing to a failure in a component when the idea was tested in principle. The design is now universal.

The first box girder bridges had a shaky start in more ways than one, because of an erroneous mental image of how they would behave; the fault lay not in them but us.

So far we have established three things about the idea in principle as demonstrated by direct testing. The first is that the right idea can be shown to be right. The second that a wrong idea can be shown to be right. The third is that it can also show a right idea to be wrong.

But there is a fourth phenomenon about judging a principle by this means, and it the most likely to happen of them all, for it concerns ideas that are a mixture of right and wrong, where truth and error are tangled up together. It is almost certainly a fact that no new invention has ever been 100% correct. Men felt the force of the wind on their faces and one day they invented the aeroplane – lifted, they imagined, by the pressure of air beneath its angled wings. The planes flew but it was the suction above the wings that was the decisive factor. Vast liners were made possible by the invention of the Michelle thrust bearing, considered a classic example of the application of mathematical theory to practice for nearly fifty years – a theory now known to be only partially true. Steel-frame buildings were invented and designed and have stood the test of time, despite the theories about their failing strength being erroneous. Their designers thought in terms of elasticity, the buildings knew better and behaved, fortunately, in a plastic manner.

Men conquered the air, the sea, and the forces of gravity and congratulated each other on their cleverness. Nature must have laughed but kindly did their bidding for the sake of the partial truth they had found out about her. And this brings us to the basic problem of direct testing which is its inability to unscramble right from wrong. To ask it to do so is to ask the impossible; because, as it cannot discuss a principle, it cannot justify one. The realistic question to ask is whether a direct test shows that the new idea is likely to be a useful one. The usefulness of a direct test is that it often can, with certain safeguards, demonstrate the usefulness of an idea. Don't ask it to tell you directly the total truth about anything, don't give it ideas above its station.

The usefulness of a new principle is best judged by its ability to predict the future. It is necessary, for instance, so to design a bridge so that you can be certain, in advance, that it won't fall down when built.

For example, the usefulness of the plastic theory of steel structures is that you can foretell the ultimate load that such structures will withstand, and its value can be measured by the accuracy of its forecast. A principle that can predict nothing is, in an engineering context, worth nothing. We can now write down the first five steps in attempting to establish the usefulness of a basic idea in principle.

1. Define the idea.
2. Predict how a particular application of it would behave.
3. Make a full-scale direct test.
4. Compare your predicted and the actual results.
5. Do it again in different circumstances.

No idea should be accepted or rejected on the basis of a single test for that would put you at the mercy of coincidences. Take a number of well spaced shots, not only to see if your shots are on the target but whether your target is in the way of your shots. Put them round the edge of your mental picture; put a few of them off the edge, where you would predict they would not work, and check. Do not be too easily satisfied.

A short time ago a company patented a new process which, in effect, was a combination of a process A, a fairly simple one, and process B, much more complicated. They would not, I think, have succeeded in patenting either except in combination with the other. They asked me to advise them on its development and showed me an impressive working machine. I asked them to demonstrate what would happen if 'A' was used without 'B', and 'B' used without 'A'. They rather scorned this suggestion but I carried it out. 'A' by itself produced results of a sort but not nearly up to the standard required. 'B' by itself produced utter chaos. By playing about with 'A' alone I ultimately found that excellent results could be produced. The sole contribution of 'B' had been to compensate for the maladjustments of 'A'. By firing intentionally off the specified target I had hit the really useful one. Rather uncritical enthusiasm, combined with a subconscious desire to invent something patentable, had masked the fact that much of their picture had been inessential and misleading.

We are now faced with the acutely difficult problem of deciding, in general, how many direct tests are necessary to establish with confidence the usefulness of an idea.

The best way of approaching this subject is to discover how the original inventive idea arose. There are three possibilities; the intuitive, the evolutionary, and the systematic. Often they overlap but it is easier to define them in isolation.

First there is the totally intuitive. The new idea suddenly bursts up, unannounced, into the inventor's imagination, often when he is relaxing and not thinking of his work at all, although he has previously concentrated on a problem and saturated his mind with its content and context. Relaxation has allowed energy to flow down into the subconscious where, unknown to the inventor, everything is buzzing with activity.* And then, unannounced, the new idea squeezes through the interface between the conscious and subconscious and erupts into the imagination. This interface transmits some ideas and blocks others.

We cannot inject inventiveness into ourselves if we lack it. But we can educate our subconscious interface to filter out intrinsically silly ideas. That is one of the great practical assets of a mind trained in engineering science; it censures the irrational, or at least some of it. The inventive but uneducated mind tends to overload its imagination with a jumble of good and bad ideas. Education, if it is wise, is an unconscious discipline which stops us making fools of ourselves. But this interface filter is evolved, not only from our formal education and informal engineering experience, but also from a welter of fears, prejudices and irrational likes and dislikes.

A surprising number of revolutionary inventions have been pro-

* 'The Design of Design', page 18.

duced by people who have been working in completely alien industrial surroundings. They have been told, or unconsciously conditioned into believing, that certain things are 'impossible' in their own field, but have no such similar inhibitions in others. A little knowledge can be a dangerous thing; a lot may be even worse.

The interface filter is both a blessing and a menace. As an inventor becomes older his purely inventive powers may decrease but his interface, if he learns wisely, will allow him to concentrate on the useful ones.

But we must also consider how our experimental technique should allow for the type of imaginative image that bursts up through the interface. Our inventive ideas may be unlimited but the form in which they can be communicated to our imagination is quite the reverse. Our picture of the new idea will only appear as three-dimensional sketches; never algebra, never symbols, only pictures.* Our subconscious is poorly educated; it can't talk, only draw. And the complexity of the drawing is strictly limited. If the invention is an involved one it can only be imagined in a series of successive pictures; like slides. And it is very difficult to be certain that the slides touch; there may be a gap; our total picture may not be complete, or else the edges may not fit.

From all this a number of practical implications emerge.

The intuitive invention of an idea in principle has probably been responsible for more progress than any other method; it should therefore be greeted with the greatest respect but also with an almost equal amount of suspicion. As we have already said, one direct test can neither prove nor disprove its usefulness; you must try pot shots over and around the target area; and always do one more than you think necessary. And when the invention is a synthesis of a series of pictorial ideas your tests must embrace the whole, not just the individual parts. Only a total interpretation can prove the overall comprehensiveness and compatibility of your mental jigsaw.

By and large, the direct test has the advantage of being convincing and the disadvantage of being too convincing. A kind neighbour once gave us three little plants to put in our garden. Their arrival coincided with my wife's departure to look after an assortment of grandchildren. A few days later my wife asked, over the phone, if I had remembered to water the new little bushes. I confessed that I had forgotten to plant them at all. I said I would do so immediately. Taking a torch and a spade I went out into the darkness of the autumn night and planted the bushes while I still

* 'The Design of Design', page 19.

remembered. When my wife came home she pointed out that I had planted them upside down, their little branches in the ground and their roots in the air. But they flourished. But on the basis of this one direct test I am not recommending you to do the same, far less take out a patent for the idea.

Safety lies only in numbers, if you can afford them. What to do if you cannot, we shall consider later. But what no one can afford is the time, money, and energy-consuming frustrations of attempting a detailed embodiment of some jealously guarded wishful thinking.

It is very difficult to find realistic accounts of how a new idea has developed but Stephen Salter has written vividly on how to extract useful energy from sea waves. His account was published in the Edinburgh University Bulletin in October 1979, from which the following are extracts.

'In September 1973 I caught 'flu. My wife said to me, with callous indifference to my misery, "Stop lying there looking sorry for yourself. Why don't you solve the energy crisis?" It seemed a good idea at the time. What she wanted was something which would provide the vast amounts needed, which would be clean and safe, would work in the winter in Scotland and would last for ever. It is a good thing for an engineer to have the design objective clearly specified.

'I thought about several things and then did a few sums about the energy in sea waves, guessing at their sizes. I was amazed at the amount of power that seemed to be available. The obvious extraction mechanism was something like a lavatory ball-cock bobbing up and down working a pump. I am fortunate to work for a University enlightened enough to provide me with my own workshop facilities. I could get hold of enough balsa wood and transistors to try some tests. It took me very little time to make a dynamometer which could measure the work done by a bobbing ball-cock. Clive Greated in the Department of Fluid Dynamics was kind enough to lend me a tank with a wavemaker in it. The ball-cock floats got out about 15 per cent of the available energy. But I found that if they were tipped so that the hinge was below the surface the extraction was much higher, about 60 per cent. It looked as if the to and fro movement was better than the more obvious up and down.

'I then tried a vertical flap. Its movements displaced water behind it to make a new wave with about 25 per cent of the energy, and the extra impedance that this produced made it harder to move, so that it reflected about 20 per cent back to the wave-maker. I got about 40 per cent out. What I wanted was a flap with no back. I made something like the British Standards kite mark. Its round rump displaced no water as it moved, and could contain a strong mounting. I worked out a shape of pump which could fit inside. Kite got out 70 per cent. Then there was Tadpole which was no better and harder to make.

'In a travelling wave in deep water each particle moves in a nearly circular orbit. At the surface the diameters of these circles are the same as the wave height, and the diameter falls off exponentially as you go down. We knew that the round back was right. If we could make the front have a shape which displaced water in proportion to the sizes of the circles at each depth, then the wave would not notice anything different. Peter and Dennis and a computer tried to grow the shape down from the surface. I had a feeling that we should grow it up from the bottom but they were quite sure.

'I went away and tried with a slide rule. By Saturday morning I had a combination of a circle and a tangent which fitted fairly closely the displacements we were after. I made it out of balsa wood and went out for lunch. But on the way I saw Peter's car parked, and realised he must be in the computer room. I went back to fetch my model, still wet with varnish, and took it in to him. When he saw me he said "It does grow from the bottom. Come and see". He had used a one-to-one scale on his printout. I put my shape on top of it. It fitted closely. We call this shape "pregnant duck" – descriptively, if not biologically, satisfactory. (Fig. 1.1) Duck can get 90 per cent at its best and is admirably compact. We can if need be afford to throw away efficiency to buy reliability or cheaper construction.'

Now if you or I (especially the latter) had succeeded in reaching this stage of the idea in principle we might well have been so pleased with ourselves that we would have sat down in a welter of self-satisfaction and assumed we had reached finality. After all, a moving duck on an immovable mounting seemed a complete answer. Fortunately, Salter was not so easily satisfied; he took 'pot shots' round the idea and found that the average efficiency was markedly

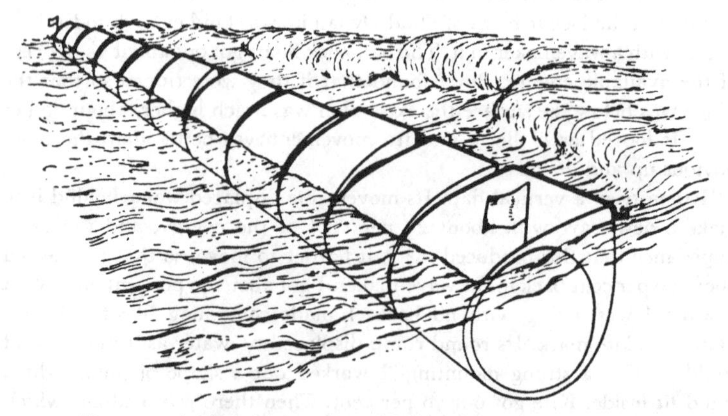

Fig. 1.1 Artist's impression of full-scale equipment at sea.

improved and the potential cost greatly reduced by allowing the duck to move backwards in the water. It is not something we would have guessed.

The next step in assessing the value of a direct test and how many there should be is to consider it in relationship to the next type of inventive thought. This is the evolutionary approach which, on the face of it, is in direct contrast to the explosive inspiration of the intuitive; but, as we shall see, the matter is rather more complicated than it appears at first sight. One difficulty is that the word evolution can mean a number of quite different things.

Used in a popular and general way, evolution means a gradual but steady forward march of progress. We can talk about the evolution of airplane propulsion from the early radial engines to the modern jet; the evolution of ancient masonry bridges to the modern suspension ones; from the diode to the transistor. In each case we visualise a steady uphill progression; things improving with time; a two-dimensional picture.

But in both the natural and exact sciences the word evolution is used in a quite different sense. This can be seen in the natural sciences, where the word originated, but seen more exactly in the exact sciences.

In its scientific sense the original biological theorem of evolution traced the progress of life from its early beginnings to the vast prehistoric monsters that once roamed the earth. But then evolution suddenly changes direction, for a little figure with a big brain, a man, appeared and a new epoch began. All scientific evolution follows this pattern. It is not two-dimensional but three; the steady rise always alternates with sudden changes of direction. In engineering science this third dimension appears equally vividly but more often.

Some years ago I felt it would be fun to have the fastest registered road car in Britain. Naturally this could not be bought, as someone else could buy one too, so it had to be designed and built specially. At that time the largest car engine available was being built in America and one was sent over. I put it in a very light aluminium bodied two-seater car. With the engine, the two-dimensional theory of the bigger the better worked beautifully. In wet weather you could start up at traffic lights on a wet road and immediately the speedometer would register over 100 mph, 95 mph of which was due to the wheel spinning, with clouds of steam rising from the rear wheels. If the car suddenly encountered a dry piece of road the acceleration was most interesting. Undoubtedly the car could move; stopping it was a more difficult problem. I concentrated on the bigger the better technique here too. The first car I ever owned had drum brakes of 5 in diameter; for my new car I made the drum brakes of so large a diameter that if there was a puncture the car went along on the drums for they were larger than the rim of the wheel. Even then their stopping power faded, generally at precisely the wrong moment. Shortly afterwards I greeted the invention of the disc brake, a fraction of the size but a great deal more effective, with considerable relief. The evolution of bigger and bigger drum brakes was a steady evolution; so is the steady improvement in disc brakes. But, in this two-dimensional sense, a disc brake is not the evolution of a drum brake; nor is the jet engine a straight-forward evolution of the piston engine, nor is the road bridge over the Forth the evolution of its railway bridge, nor the diode the father of the transistor. In each case the arrival of a third dimension is of a different kind from the other two, for it is sudden and creative.

If someone tells you that he has evolved a new idea in principle ask him how. If it was a two-dimensional process, and he can convince you this is so, direct tests of the principle are unnecessary, as no new idea in principle has been introduced, so go straight to the first embodiment of it. If it is a third dimensional creative step, treat it exactly as an intuitive one. But if he tells you that he has evolved by linear argument a new idea in principle, be highly sceptical. In the world of science it amounts to a self-contradiction.

There is, I think, a particularly engineering meaning to the word 'evolution'. It is not 'the survival of the fittest' but 'non-survival of the unfittest'. New ideas can, and often are, developed in just this way. In a sense they invent themselves by shedding liabilities in an *ad hoc* way.

The development of steel for the reinforcement of concrete is an example of this. Forty years ago, mild-steel bars were almost universally used for this purpose. To prevent them slipping through the concrete their ends were bent into hooks (as in Fig. 1.2). As the purpose of the reinforcement is to collect and redistribute forces, it might be a good idea, it was felt, if the bars hung on to the concrete along their whole length. The obvious thing to do was to take two adjacent bars and twist them, in the same way that ends of wire can

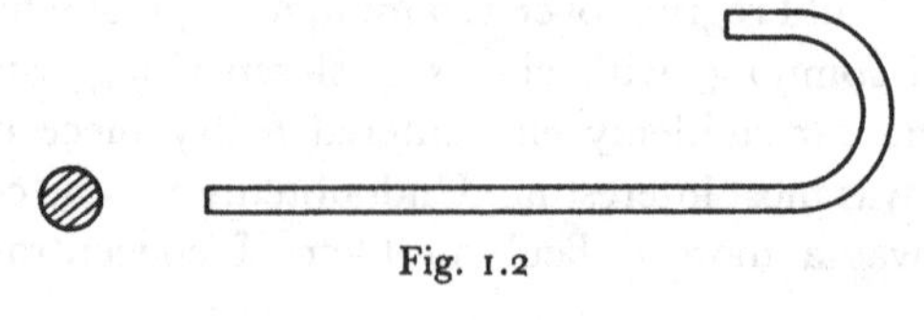

Fig. 1.2

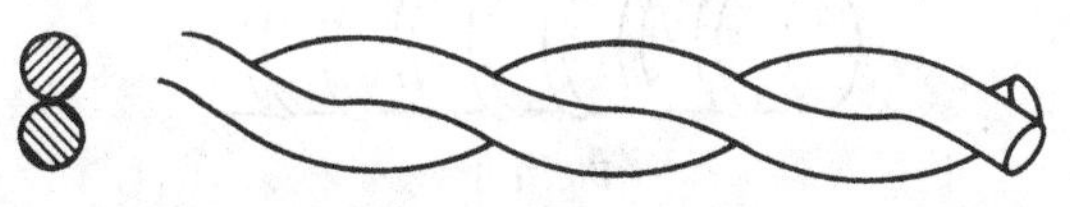

Fig. 1.3

be twisted together. When this was done (as in Fig. 1.3) everyone was delighted to discover that the bending and stretching of the steel had removed the normal yield characteristic of the mild steel, and so its effective strength in concrete was some 40% higher. This was nice of it. No one, as far as I know, expected this to happen in advance but everyone was suitably grateful. The configuration had, it was soon found, a liability. It was no good at absorbing compressive forces; it just started to unwind. To shed this liability it was easiest to roll the steel into a figure-of-eight section to start with, and then twist that (as in Fig. 1.4). It could then take compressive forces too. Next the steel rolling mills pointed out that it was an expensive section to roll; can't something simpler like a square bar be used (as in Fig. 1.5). So a square bar was tried out and behaved better in every way than that of Fig. 1.4. Another happy surprise all round. This design was, and is, used extensively, but it carried with it one liability. Its flat spiral surface hung well on to the concrete provided the section of the steel was not too large. As the size increased so did the surface area of the bar, but the stress within it to be transferred to the concrete went up with the *square* of the size. This stress, combined with the wedge-shaped spiral, tended to split the concrete. This liability was shed, in turn, by rolling cross ribs on the bar as shown in Fig. 1.6. Here seemed to be final development, until very recently, when highly stressed, corrosion-prone, concrete rigs were being built for the North Sea oil industry. Up to the present, fatigue loads in concrete reinforcement have been insignificant and

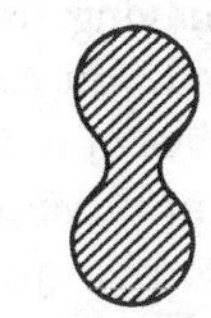

Fig. 1.4

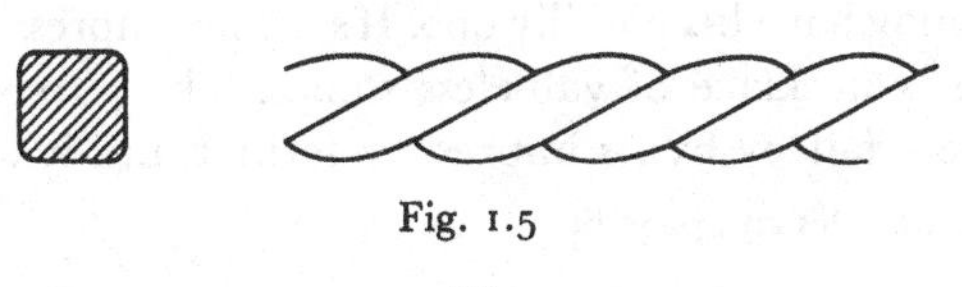

Fig. 1.5

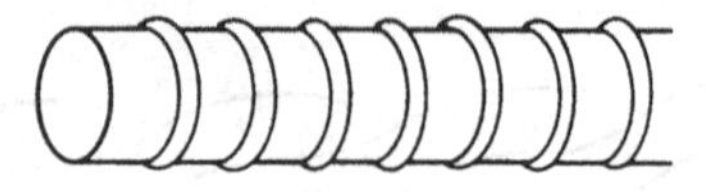

Fig. 1.6

rightly ignored, but now we have to watch out, and redesign the cross ribs so that they do not encourage local cracking or stress concentrations. And so it will go on. A self-inventive evolution where the unfit characteristics are jettisoned in turn.

The disadvantage of this pragmatic approach is that it follows events rather than leading them. The beneficial effect of strain hardening on the reinforcement was a lucky break, but you cannot assume that you will always be so fortunate. Concentrate on the two dimensional and sometimes you may find the creative third dimension thrown in free of charge, so to speak.

The third way in which a new idea may be created is through some systematic method of thinking. There are many of them. But they can be roughly lumped together into a few family groups. At the moment we are not concerned with their relative merits as sources of new ideas in principle; we merely wish to answer the question 'Can we safely reduce the number of direct tests on an idea because it is systematically and not intuitively inspired?'

The most systematic method of all is a geometrical permutation. For instance, having decided that a net conveyor is best for a continuous laundry washing and ironing machine, you can draw down in sequence all the main geometrical possibilities.* As in the ringing of a peal of church bells it can be formally done without making decisions.

The Wankel Engine in its most usual form is one particular geometrical application of the principle that has over 800 geometrical forms. The characteristic of the totally geometric is that you ultimately trace out a circle, having travelled round all the possibilities. Wherever you start, and you can start anywhere, you will come back to it again. On the face of it, the system appears to be less open to error than the totally intuitive but this is, I think, a fallacy. You will notice that the geometrical juggling can only begin after you have something to juggle. First you must decide on a net conveyor before you can consider net conveyors. And this original assumption may have been a silly one. If so, your impressive sequence of options is a necklace of valueless stones. The purely geometric may disguise a fallacy by its impressive format, but it will not pre-

* 'The Selection of Design', page 69.

vent one. It is only an intuitive idea in a party dress. Treat it as such and do not reduce your number of direct tests because of its apparent glamour.

The geometrical has, however, one outstanding asset; in ensuring that all possible forms of an idea are displayed it allows you to select which form of the idea is most likely to be useful and we will return to this later. Within the context of a creative principle it is a non-event.

Next we come to many different ideas that can be classed as logical systems. They all require value judgments and the filling in of certain patterns of thought, and many people find them invaluable as a design tool. And a systematic pedigree here has one great value; it can and, in fact, should ensure that the solution, good or bad, is one that fits the problem. It may not solve it well, but it should fit it well, for most logical systems include a careful analysis of the restraints involved. Some start with the considerations of them, others feed them in sequence later;* all define them. As the ideas in principle so produced have had a partially disciplined background, you may well be able to reduce the number of direct tests because of it, or, perhaps more accurately, you will be able to select more easily the precise form that these should take; less hit and miss and more hit.

In practice I am inclined to believe that these systematic patterns of thought are often most usefully used backwards. Inventors have the intutitive idea first, and then, in an attempt to enlist financial support or interest in it, sit down and plot out a sytematic scheme to show how logical it all was. This may, of course, have the fringe value of showing the inventor how his original idea ought to be modified somewhat, but otherwise it is little more than a confidence booster. As in G. K. Chesterton's poem, you can go to Birmingham by way of Beachy Head; but you may think it more respectable to imply that you went there direct.

But before leaving this subject it is essential to sound a warning about systematic methods in general for devising or confirming new ideas in principle. There is a distinct danger, as there is in other realms of design, of confusing a conclusion with a disguised assumption. To make a systematic method possible at all a number of conditions, parameters and so on, have to be written down. Some are clear and objective, others you must guess meanwhile. Unless you put something down the system can't be followed. You then follow it all through and reach a conclusion. So far so good, but you must not stop there. You must go back and find to what extent your

* 'The Selection of Design', page. 15.

final verdict is dependent on those original, non-objective assumptions. A guess dressed up in a fashionable systematic garb is still only a guess: a sheep in wolf's clothing.*

Summarising briefly, we can say that the direct test, for justifying a new idea in principle, is deaf and dumb but not lame. It shows the usefulness of an idea through a series of steps. But do try to refrain, if you can, from pushing behind or running in front.

During and after the last war I was in charge of a research organisation concerned with reinforced concrete. Much less was known about the subject in those days, especially by me. One week we were doing a long series of tests on the failing loads of beams reinforced with round, mild-steel bars. When the stress in the steel reached that of its yield point, the reinforcement suddenly increased in length with a corresponding decrease in diameter. This decrease allowed the concrete to slip back along the steel and so caused such wide cracks that the beam failed. You could almost see it all happening. Discussing this with a colleague, I discovered that his mental picture of what was happening was as vivid as my own, but quite different. He visualised that the wide cracks were caused by the elongation of the steel directly and, as the concrete hung on to it, wide cracks were formed.

The usefulness of the tests was that they confirmed a prediction that you must not assume a higher stress in the steel than its yield point. It did not comment on which mental picture was correct. We were both reading into it our own ideas. Thinking we were believing it, we were only believing ourselves. The direct test is often maddeningly inscrutable, but you must learn to accept the fact. It just won't answer leading questions.

So far we have been entirely concerned with the relevance of full-scale direct testing to our inventive picture images, the concrete confirmation of the abstract. But there is also the reciprocal of this to be considered; where observation takes the place of imagination; the objective from the subjective. Perhaps this process is best defined by illustrating it.

I was once engaged in studying a continuous production line for specialised steel. I had a clear idea of what was happening in general but needed confirmation of some details before I could redesign the line for faster working. While watching closely what was happening a fuse suddenly blew and the machine immediately stopped, leaving the steel in a partially processed condition. Seeing what it looked like, I realised that my previous mental picture was sadly lacking.

* 'The Design of Design', page 78.

I now saw what was really going on and redesigned the line. Today these patented machines are working in many different continents, embodying the new picture. The fact that many people, much more intelligent than myself, had apparently never realised what was actually happening was due to the fact that a fuse had never blown under their noses. The direct test had led to the abstract theory. The cart had come in front of the horse.

Now, on the face of it, this would appear to have given us a flying start for our research; would we be justified in reducing the number of direct tests for this reason? I think the answer must be 'No'. As ever, we have been told 'What' – 'Why' we must supply for ourselves, and we may still get it wrong. Our observation-based invention indicates an idea, and encourages its further investigation, but does not fundamentally reduce the possibility of human error. Once again, don't ever depend on a single test. The early Rolls-Royce cars were designed on the principle that 'if you think one will be sufficient always use two'. Always do one more test than you think necessary.

In addition to inventiveness and observation we have a third possibility for the initiation of new ideas in principle, and today it is probably the most potentially useful of them all, i.e. discovery.

We earlier pictured the engineer as the scientist who stopped at home while all the others trotted in and out of the jungle. Imagine one such natural scientist returning to his beach-head home, exhausted but excited, hobbling in with the help of a stick. He throws the stick down, demands food and tells us all about his discoveries. We listen, fascinated, and the engineer, trying to tidy things up a bit, picks up the discarded walking stick. After fidgeting with it for a while, he interrupts the returned traveller to ask where it was found. The material of which it is made, he says, might be useful to use in a bridge, a bed, or a boomerang. The scientist discovers a new type of material or energy and the engineer discovers a new use for it. Lasers, silicon chips, superconductivity, magnetic alloys and photosynthesis are random examples. This process is best defined in the form of a contrast. The engineer is concerned to travel from the abstract to the concrete. He begins with an idea and ends with an object. He journeys from theory to practice. The scientist's job is the precise opposite. He explores nature with his telescopes or microscopes, or much more sophisticated techniques, and feeds into a computer what he finds or sees in an attempt to define mathematically its significance and relationships. He travels from the real to the symbolic, from the concrete to the abstract. The scientist and the engineer are the mirror image of each other. When an engineer wishes for a new material or energy source from nature he must go to nature's explorer, and their combined efforts can be pictured as a rainbow journey from the concrete up into the abstract and down again into the concrete. I have written elsewhere* of the general significance of this but we are only here concerned with a certain aspect of it. We want to know if any particular safeguards are necessary, when, as in the case of energy or materials, they are the outcome of the boomerang from concrete to concrete via the abstract. I think there are several which apply to these special cases, safeguards which are apparently easy to forget.

The first complication lies in categories of thought. The biological scientist is engaged in a non-exact science, the engineering scientist in an exact one. The latter must design exactly using a non-exact material or energy. He has an idea or principle that he thinks might succeed but everything depends on whether the new scientific discovery can be defined in a sufficiently exact way to be absorbed into or manipulated by the essential mathematical and design disciplines that alone can make it an asset to the community. And the key question is exactly how much do we need to know. I think we must distinguish here between our two types of discovery, i.e. new types of energy and new types of materials.

You will never know all there is to know about any source of energy; but you must know sufficient to exploit it safely and economically. Steam locomotives chuffed happily up and down the land for a century before anyone began to realise exactly what made them puff. It is sufficient to train up your new-born energy in the way it should go; don't try to psychoanalyse it first. Almost certainly direct full scale tests will be initially the least efficient way of researching into a new source of energy. It will be at once too expensive and too dangerous. If you want to develop a steam locomotive, start, as Stephenson did, with the family kettle. This statement may appear too obvious to be argued, and it certainly is after you have ignored it; not always before.

The first piece of serious research I attempted was when, at the age of about seven, I had successfully pressurised my parents into giving me a toy 'Meccano' electric motor for Christmas. This was operated by a 6-volt dry battery which soon lost interest in the proceedings; my interest, on the other hand, was rapidly increasing. I decided that if I connected the terminals of the motor direct to the 200-volt electric light mains it would be illuminating. Actually the

* 'The Science of Design', page 1.

reverse happened, as half the lights in the house immediately went out. Hugging myself with glee, I followed my father into the cellar while he muttered about how unreliable fuses were made nowadays. Those which had not already been replaced with thick garden wire or a nail were rapidly brought into line. I felt that perhaps my first experiment was a little crude and I had better find a way of 'slowing the electricity down a bit' as I thought.

As soon as possible I went down to the ironmongers and bought as much bare thin wire as I could afford. I searched about to find something large enough to wind it all around, suitably spaced, finally decided that the chimney stack was the only thing big enough and wound the wire around it. I also built my electric motor into a model car which it drove through a doubtful reduction gear. It was quite a moment when, with my electrically driven car pointing across the attic floor attached to many feet of trailing flex, I turned on the light switch. The result was enthralling. Looking like a highly mobile Catherine-wheel firework, the little car bolted across the attic, tried to climb the opposite wall, fell on its back, and short-circuited the leads. There were sparks everywhere and the chimney stack must have been warming up the neighbourhood. There was never a dull moment. Nor will you have one if you try the direct full energy testing of a new source. And you won't have time to learn much either.

The real danger here is that you will dismiss the whole subject as being too obvious to waste time talking about; but it is unwise to assume this. The problem may well be that your mental image of how the new energy source is going to work will not coincide in all details with how it actually will.

Sir Frank Whittle describes his experience with the first jet engine as follows.*

'The very first attempt to start was successful in that the engine ran under its own power, but it accelerated out of control up to about half its designed speed. This happened several times and was very alarming; in the early days the people in the vicinity did more running than the engine.'

Equally unexpected was the failure of an experimental windmill which suddenly tossed one of its 8-ton blades 750 ft.†

In general, the problem lies in the acute pressures which impinge on an inventor: his natural anxiety to see if his ideas will work; the rapidly vanishing stock of money and so on, are all on the side of

* 'Engineering Progress through Trouble', page 11. Inst. Mech. Engineers.

† 'Power From The Wind', page 131. Palmer, Cosslett and Putnam.

starting up the sources quickly. Enthusiasm for stopping it has less self-induced enthusiasm behind it, and may be given less thought.

The only exceptions to this rule occur where the discovery of a new energy source is itself linked to size. Apart from nuclear engineering and micro-electronics, I do not think this happens very often; in my own experience only once.

This was in the development of the sonic pile drivers which are used on building sites. One of the orthodox methods of inserting concrete piles into the ground is by first driving in, by a hammering weight, a steel tube with a plate masking the bottom end of it. When it has been bashed possibly 30 metres into the ground, steel reinforcement is dropped in through its open top end, followed by liquid concrete. The surrounding steel tube is then picked up out of the ground, leaving the reinforced concrete column to set in the hole. The normal diameter of the steel tube is about half a metre and it may take a whole day or more of banging to persuade it to go into its full depth. The new idea was an alternative source of energy which might be quicker and quieter. Science has long since established that there is such a thing as a resonant frequency of vibration. Why not put the steel tube into a state of resonant high frequency vibration and its lower end might dance its way into the ground? Might or, on the other hand, might not. The resonant frequency depends on the length of the steel tube so a full-scale test was the only valid one. Little bits of tube going into little bits of earth could not provide a realistic answer. The full scale test equipment was far from cheap; up to 1000 hp, remotely controlled, had to be generated and converted into high frequency vibratory energy over 30 metres up in the air, and be lowered in a controlled manner to follow down the tube to which it must be attached by a quickly releasable clamp. The production of all this power into solely vibratory energy was without engineering precedent, as was some of the ancillary equipment. If something went wrong with the mechanical arrangements it was sometimes difficult afterwards to determine precisely what had happened. The equipment not only suddenly overheated, not only became red-hot and then white-hot, but melted and ran down the tube like strawberry jam.

Avoid direct full-scale tests of new energy like the plague and only be argued into agreeing to them if there is an inbuilt and unavoidable scale factor. If you want a rocket to go to the moon in, don't start your experiments with a Saturn-sized one. Fortunately, in this instance, prudence and economy pull the same way.

The efficiency of the direct testing of newly discovered materials is most difficult to deal with in a general way, for we have no reason to believe that such materials will conform to the overall pictures of the past. You cannot decide about discoveries until they are discovered. You must therefore start at the other end. Unable to anticipate what he will find we can, to some extent, determine what an engineer is looking for. He has an open mind but not a blank one. To be proved potentially useful a new material must be tested for three vital characteristics, consistency, stability and cost. I am starting off with consistency not because it is usually the most critical feature but because it is the most usually forgotten or ignored. A material that is inconsistent is not a material at all; it is a collection of them and a gamble which sample comes out of the hat.

I first met this phenomena when I was asked to sort out a production line intended to exploit a new material. Amongst other requirements this material had to be processed at a critical temperature, with a maximum variation of $\pm 1\frac{1}{2}$ °C. Due to a non-measurable non-uniformity in the material itself this critical temperature needed to wander over a range of ± 3 °C. It was therefore an intrinsic impossibility to control the temperature for all variations in input.

In the original experiments into the new material it had been found that all the samples could be processed satisfactorily; but the experiments failed to notice, or rather failed to appreciate, the significance of the unavoidable and unnoticeable slight variations in the critical temperature. More haste, less speed; which means more money down the drain. Again, the chief chemist of a firm I was advising once came to me with an idea of a new material in principle which certainly sounded hopeful. I arranged direct full-scale tests but using up as little material as possible in the circumstances. I needed to establish the degree of consistency of the material before doing any actual production runs.

Meanwhile I had to visit another factory in the group, and then I caught influenza and it was a month all told before I returned to ask how the consistency tests had turned out. The chief chemist said that these had been stopped by a high-up personage in the organisation on the grounds that 'we can't wait for that', 'time isn't on our side', 'we must get a move on'. A day's production run had been organised. The material turned out to be decidedly non-uniform, the final product even worse, and all of it had to be scrapped. The chief chemist was ticked off, and being fed up, soon left, which was a great pity for the company. No-one would consider having any further experiments along the lines he had suggested, which I still think were most promising. I often wonder why non-technical personnel frequently seems to assume that all materials are consistent when none of them are. All vary between limits and the first essential is to establish what those limits are.

On another occasion I had a front-row view of this blind spot and its results in a firm negotiating to manufacture under licence a process exploiting a new material developed in the USA.

I was engaged on other problems, and this one was quite literally none of my business, but I could not help hearing what was going on. The Americans, who behaved in an exemplary manner throughout, were pointing out that the new material, as dug out of the ground initially, was subject to variations that proved embarrassing in production and they could not say how the European equivalent would behave. All this was swept aside with such phrases as 'our chemists will soon sort that out' and so on. A licence was negotiated, a completely new and expensive production line built, with complicated heating arrangements, and nothing saleable was ever made in any worthwhile quantity; the raw material was too inconsistent. The only consistency about ignoring inconsistency is that you lose a great deal of money. You must do direct tests on a large enough scale to allow the natural variables to frolic about in an uninhibited way, and so disclose their true range.

The next criterion for assessing a material is its stability. 'Statics' is sometimes a misleading term for a subject; it implies that materials don't move about, but they all do. And the vital thing to establish is where they move off to. Don't worry about baptismal names, teenage behaviour is what counts. Materials either grow up or grow down, you cannot usually halt the process. If a vital part of a bridge is failing by fatigue all you can do is to prepare to run. It is a paradox that the dynamic, say a new turbine generator, can be turned off and become stationary for inspection; while all materials have internal stability drifts that you can't turn off. You can often make the dynamic static but you can never stop the nominally static being permanently dynamic. So if someone rushes up to you and claims marvellous possibilities for a newly discovered material, don't ask him what it is good at, ask him what it is bad at; that is more important. Even up to a few decades ago the instability of materials was rarely appreciated. They were classified by their ultimate failing load in tension as measured in force per unit area. We know that this figure is, in general, largely irrelevant. This is not only because in most cases the distortion arising in the structure failure makes the initial calculations meaningless, but more importantly, the influence of corrosion, creep, fatigue, size and surface finish, and several of them often in combination, cause failure well before the stress a laboratory tested specimen would suggest.

Direct full-scale tests provide not only the best but usually the only realistic method; but they have one great liability. As many of the possible instabilities, such as fatigue, creep and corrosion are actively linked to a time element, full-scale testing means full-scale time testing. Years, in fact. And who is going to wait that long? It is an acute dilemma. In practice we often have to do some guessing and engineering history abounds in examples of unlucky ones; and some lucky ones too.

An expert is, in general, someone who knows more and more about less and less. But an expert in materials generally knows less and less about more and more. Often the best you can do in the direct full scale testing of a new material in principle is to establish how it behaves under the various possible conditions which are not time-interlocked. We shall have to return to this subject again as it is an inbuilt hazard through most stages of development.

Finally, we come down to the question of cost. In engineering we always do. It is part of the enthralling challenge of the subject.

All materials, new or old, are supplied by the universe free of charge. Nature never sends in a bill. The cost of digging up, pumping about, or sucking them in, is what we ourselves pay for helping ourselves. And, moreover, we rarely find our materials sitting there in the precise state we need for their immediate engineering use. They must be refined, processed, purified, or refashioned to suit. All this is likely to be expensive, too; and so is the energy consumed.

If a new material, in principle, imported from the discoveries of the natural scientist, is made in small quantities by hand, the ratio of hands to usable material results in prohibitive cost. Make it in large quantities in vast machines with few hands and the interest and depreciation charges on the huge capital outlay are also prohibitive, unless the new material is decisively better or an essential precondition for the viability of something else.

Next, for inventions in general, there is that form of direct testing that is the partnership of measurements and mathematics. Here you save time and money by testing isolated components or assemblies and fill in the gaps with calculations. It will only work safely if both are direct, in the sense that neither have sandwiched in those indirect subjective assumptions we have already discussed. Some 'practical' people feel, and even say, that theory and practice don't agree and are blandly sceptical of mathematical logic. Actually we live in a mathematical world. The difficulty about some self-styled 'practical' people is that they are so impractical. There is more in the universe than meets the eye.

Imagine an engineering research worker bicycling back to his home at night. Bicycling because he is underpaid or overweight, or both. His bicycle, happily obeying the laws of Newton, becomes progressively harder to pedal uphill and he stops for a rest. He leans on his bicycle and looks up at the stars. He knows he cannot realistically imagine what this large-scale world is like. He can enjoy the impact of its beauty; he cannot picture the curved space of Einstein's relativity. The stars in their courses are far above his head in more ways than one.

Then he looks down on to the pedals of his bicycle. They are solid and obstinately hard to move, unescapably real. And yet he knows them to be but a consortium of random, unimaginable and immeasurable particles of energy. The quantum mechanics of Eddington seem even more surprising and unreal than the giddy heights of Einstein. A shiver recalls him to his surroundings and he mounts his Newtonian bicycle and pedals home to the comforting reality of a drink, a fire and carpet slippers. This is a picture of how the world appears relative to him or us. Human scale reality sandwiched between large and small scale dream worlds.

We live with the normality of Newton and his laws. The mathematical relations of Einstein and the quantities of Eddington cannot, we feel, leave the paper they are written on. We feel our sense-reacting world is inescapably concrete; sandwiched between abstract theories.

But what of a spectator viewing our universe at a distance? He would see a large-scale world, dominated by relationships, touching a small-scale world of quantities. Touching, but more than touching, overlapping slightly, creating a wafer-thin no-man's-land where all men live – the special freakish world of Newton, the world where relationships and quantities, algebra and arithmetic, are the fabric of reality and the more we know about them the more practical we shall become.

If you can't do mathematics, go to someone who can; we all have a ceiling to our mathematical abilities; whether it is high or low is a matter of luck or hard luck. It has no relationship, either real or implied, with our ability as a creative artist in engineering science. If you are good at both, fine; if you have to choose one, be an inventor every time, for largely on him will depend the prosperity of his community and country. But we must return to our main theme, the down-to-earth realism of direct tests, interspaced with direct mathematics; for it has a further asset not so far mentioned. It not only tells you if a thing will work or not, but also how far away it is from not working, or the margin of safety if it does. Moreover it may stop you from attempting the intrinsically impossible,* for the laws of physics are not open to negotiation.

Finally it must be emphasised that the direct testing of a new idea in principle is a limited objective; see that you limit yourself to it. An idea in principle is one thing; the first embodiment of it is quite another; something halfway between the two is neither of them; its object will be confused and its outcome confusing. The difficulty lies in the fact that, in some circumstances, it is a most tempting thing to do.

A direct full-scale test can act, as we have seen, but not talk. The inventor, like a ventriloquist, must do all the talking. His test apparatus is the doll, to be dressed up as convincingly as possible and jump about suitably to impress the audience. The onlooker is often the man with the money but not necessarily equipped with research insight. Everything must be done to convince him. He wants to 'see it work'. And so the idea in principle must look as much as possible like a full embodiment of it, with all the attendant clutter of 'practical' accessories. It must not only work but be seen to work by the layman. Never show children, fools or bankers a job half done is the thought behind it all. But the banker may well be a great deal more astute than you think he is, and by trying to fake up a doll to look like a fully embodied human being you are deceiving nobody, except possibly yourself.

But, in practice, a more self-evident disaster will very likely happen, for a demonstration loosely sighted halfway between an idea and its embodiment may well come to a sudden end because of the collapse or seizing up or falling off of those unnecessary and hasty embellishments you added for public relation purposes. This will leave you embarrassed and back-tracking, trying to explain that the idea in principle was not shackled to this particular rogue component. The basic difficulty is that direct tests, both of an idea and its embodiment, are life-size and you can easily drift from one towards the other and end up falling down between the two. If your objectives are fuzzy, your conclusions are likely to be so too.

But it is easy to underestimate these practical pressures and the commercial ballyhoo that often force an engineer, against his better judgement, to assist in dressing up a doll and calling it a genius. It

* 'The Design of Design', page 5.

is going on all the time but I must be tactful in giving illustrations.

Recently, a friend of mine successfully completed a full scale test of a new idea in principle. The apparatus was then seized on, dressed up with pseudo-impressive functional frills, some of them chromium-plated, and shown joyfully at a large exhibition. Nearly a million pounds of orders for it were pressed into the delighted salesman's hands. Alas, all the orders had to be sent back. Now, months later, the machine has been through both its first and final embodiment stages, looks quite different and works beautifully; but orders are inevitably now hard to come by. More haste, less speed.

The earliest example of this kind of disaster I remember seeing when I was about five years old. It was during the First World War and I had been taken on to Wimbledon Common in South London to watch a fund-raising demonstration. A new trench mortar was to be shown in action. Mounted on a tripod its purpose was to lob a small shell from a front trench into the enemy's one, with a range of about 30–50 metres. The crowd gathered to watch a dummy shell being fired; a band played, speeches were shouted; I squeezed through to the front row, and finally, an officer let off the mortar. Instantly the tripod fell flat on its face, there was a loud bang and the shell went 2½ miles. It was an experimental device dressed up as a final one. This I did not know until my parents read the papers at breakfast the next day; what delighted me then was the universal consternation among the uniformed officials, and a little khaki-clad figure madly winding the starting handle of a spindly-wheeled dark green car, which later chugged off in the supposed direction of the departed shell. I felt that engineering was an interesting game; I still do.

Perhaps it would be helpful to summarise the contents of this chapter in the form of a table (Table 1.1).

Table 1.1. Are multiple direct tests likely to be efficient in finding out the usefulness of new ideas in principle?

What happens	Yes
Why it happens	No
New energy	No
The consistency of a new material	Yes
The stability of a new material	Yes
The cost of a new material	No

2

The basic idea—indirect tests

The indirect test is not an extension of the direct, but its inverse. The direct test had practice but no observable principles. The indirect is all principles but no observable practice. The direct was active but dumb; the indirect chatters away but does nothing. The ventriloquist and his doll have swopped jobs.

Take an almost ludicrously simple example and imagine you have a new idea that involves a weight suspended on a tension spring. The weight, once deflected, moves up and down. You want to find the time cycle of this oscillation; so you measure it directly with a stop-watch.

But you could find the same answer indirectly. If you found out, in principle, how much the spring extended per unit mass, you could calculate the natural frequency with a pocket calculator. But let us assume you use a slide rule. The whole problem would be now represented indirectly by distances on the scales of your slide rule, and the answer would also be represented by a distance on it, which would in turn have to be interpreted back into terms of real time. The two words 'represent' and 'interpret' are the twin towers from which all indirect tests are hung.

The first advantage of the indirect test is that it can escape from the dictatorship of time. Our lives are inevitably and unavoidably dominated by the old man with a scythe, but an indirect test can answer back. It can save time, stop it, manipulate it or reverse it.

If you want to discover how the periodic frequency of your weight and spring is affected by changing the mass of the weight, you will find the slide rule much quicker than the stop-watch. And when you have the answer you can sit and look at it. Time and tide wait for no man, but both can be frozen stiff in an indirect context. And if you want to find the exact weight to give a certain frequency your slide rule will be incomparably quicker than weight swopping.

The realm of the direct test is that of a concrete detailed edifice; an indirect one is by abstract figures and logical principles to be interpreted back from their representation and manipulation within a more servile medium. And it is often a much more versatile medium too. Returning to our weight bobbing up and down on a spring, you would find a direct experimental measurement of its acceleration at any moment a complicated business. Indirectly, on a slide rule, you could reel off the figure in a moment. Juggling with abstract figures and principles is often much more rewarding than building with the concrete bricks of practice. And often it is the only practical way in which it can be done. As we have already seen, if your new idea in principle depends on the interlocking relationship of more than two widely varying quantities, the number of direct tests that become necessary may well be prohibitive. The difficulty often is that you cannot tell how many you will have to do until you have done them. And by then the money may have all gone. And even with two variables you may get lost if there is some interaction or feedback between them. The indirect method is therefore not only attractive but often essential, and the most popular form of it lies in a computer, of which there are three types – the analogue, the digital and the hybrid.

The analogue computer uses an electronic medium, where quantities are usually represented by voltages and their relationships by electronic devices selected and arranged like dominoes in a suitable pattern. It is an electrical picture representing a physical one. As we have seen, a new idea in principle often originates in a mental picture of a physical arrangement of it. The direct test deals with this physical entity as such. The analogue computer copies it on to an electronic canvas. The same principles are mirrored in a different medium. This is why an analogue computer is so attractive. Our creative imagination is at home with principles and relationships; so is this computer. We think in analogies; it works in them. We are birds of the same feather.

The digital computer thinks in digits. In inventing we don't. We can use it for analysing our ideas but we can't see or adjust what is going on so readily. It chatters arithmetic when we want to talk algebra. The analogue computer pictures quantities in variable voltages, the digital in electrical pulses; the hybrid starts with the first and then transposes it into the second at a suitable point. It intends to make the best of both worlds; so that we can start with algebra and end up with highly accurate arithmetic, and a retentive memory.

Our intuitive invention, inevitably idealised and simplified to squeeze on to the small canvas of our imagination, asks to be told if it will work or not. The indirect test, the analogue one, particularly, tells us not only what will happen but why. We know why in advance; we've built it in. The electronic 'what' we can interpret back into reality. And if our 'what' is disappointing we can modify our 'why'. We have logic and flexibility at our finger-tips, a model imagination.

This can be illustrated by a fictitious example. You are travelling, we will suppose, in a tightly packed railway carriage. Presently, to your relief, the train stops at a popular station and nearly all the passengers scramble out. All, in fact, except a mother and her small boy. The carriage, lightened of much of its burden, rocks, jerks, and sways along. A few minutes later the small boy announces, in a loud voice, that he is going to be sick. Your first reaction may be annoyance, quickly followed by sympathy for the embarrassed mother and especially for the little boy. It's not his fault he feels so awful; it is the fault of the individual who designed the suspension system of the carriage. And there suddenly flashes into your mind a new idea for a better one. Suppose that to the existing compression springs there were added variable rate shock absorbers, which acted with increasing severity as the deflection of the springs decreased. The lighter the load, the greater the damping. This might stop the swaying about (remember that this is a fairy story).

Your imagined picture of this suspension system, in principle, and components in some detail, can be fairly clearly visualised, but once you start trying to imagine what would actually happen when it all started acting and interacting you immediately get confused. Even if you built a full-size test rig you would be faced with deciding in advance what forces are likely to occur in the various components, and how the components should be rated to resist them, and you end up by having to guess values galore. Your idea might be right, but your guesses wrong. You would be most likely doomed to expense, frustration and inconclusiveness.

Your only hope is to exploit the indirect test, preferably on an analogue computer. Here the various movements would be represented by voltages which you could modify or multiply, their interaction would be represented by electronic packages and the resultant deflection of the small boy pictured on an oscillograph. You could play about with quantities, plug in differing relationships and immediately detect their effect locally or totally. You could soon find if your ideas had any promise in principle. And if they had not, you would be certain that they had not. You would be spared the widely expensive 'one more try' of the full-scale test which is often the famous last phrase before insolvency.

This introduces us to a further asset of the analogue computer. As it can tell why an idea works, it can also suggest why it won't. Having shown you, probably, that your railway-carriage suspension system is not going to do the boy's inside a world of good, it may also help you to evolve one that will. This third dimension of evolution may be a largely hit-and-miss question of substituting different electronic packages. You will need quite a deal of luck to achieve much in a completely arbitrary way but semi-intuitive guesswork may be rewarded. At least you can have a go; fairly cheaply.

But there is a vitally necessary warning that must never be forgotten in computer-based indirect research. Of necessity all quantities and processes are simplified and idealised. Unfortunately we do not live in a simple or ideal world. Interpreting back into it an idealised picture may not be ideal. Machines are often not as simple-minded as we are.

The use of a digital computer which can be regarded as a pocket calculator with a memory does not fundamentally affect the situation. Anything an analogue computer can do when dealing with a new idea, so, in general, can a digital one but in a still more indirect way. It deals in digits; but it was waves and not digits that made the little boy sick. Moreover, as our subconscious creative ideas can never be directly conveyed by symbols, such as digits, they must translate into them. The obvious asset of a digital computer lies in its almost instantaneous calculations and fantastic accuracy, but both may be unnecessary in the realm of ideas in principle. It also has the physiological danger of implying that a value is as specific as its multifigure digital representation. It may only be a highly accurate way of conveying an inaccuracy.

In most engineering problems this speed, accuracy, versatile programming, and memory store of the digital make the analogue computer look elementary and awkward; but for chasing the will-o'-the-wisp of a new idea the analogue computer is often the better. If you know or guess something definite the digital computer will tell you a great deal more; when you know next to nothing the analogue may well team up better with your pictorial imagination, but it is unwise to generalise too emphatically. All the same it may be easier to catch a butterfly in a butterfly net rather than shoot it down with a machine-gun.

It must be constantly borne in mind that the computer world is a world of fairy stories, but do not forget that fairy stories may well picture and highlight reality in a significant way.

Two of the most famous fairy stories were written by a mathematician, Lewis Carroll. One was called 'Alice Through the Looking Glass' and pictures, symbolically, the working of an analogue computer nearly a century before one existed. In it, Alice, a small girl, succeeded in climbing through a mirror and exploring the world on the other side of it. All the characters of this side were to be found on the far side but there they had to obey the author's sense of fun and imagery. He decided, in advance, that the movements of each must correspond to a game of chess. Their relationships to each other were analogous to the real world, but those to time and space he could vary. 'Now, here, you see, it takes all the running you can do to keep in the same place', remarks one of the characters. By transporting the real into the looking-glass world the author vividly interprets what happens on this side. When someone says 'Jam tomorrow but never jam today', we all know what they mean, especially sometimes those in charge of development programmes. I do not think that Lewis Carroll consciously knew what he was picturing but, at least, he made one character a puzzled inventor, and an oscillograph screen is uncommonly like Alice's looking glass.

The other book is 'Alice in Wonderland' where she dreams herself into a world where both sense and nonsense have equal standing. A character called The Mad Hatter logically and digitally computes the results of his eccentric ideas. One book is full of analogies, the other of symbols, and we too can compute in a 'looking glass' or 'Wonderland' way. In either we escape from the real world into an imaginary one. In both fairy stories and computers we may find isolated and emphasised those key principles or ideas that are normally hidden beneath the torrent of events in our moving world.

3

The basic idea—combined direct and indirect research

Possibly the best way to introduce this principle is by considering another imaginary case. You have, for example, a marvellous new idea about how to design a pylon for carrying overhead electricity cables. It involves an ingenious structure built up by welding together elliptically sectioned steel tubes. This being a fairy story, tubes of this type are readily available at a competitive cost.

One of the principal requirements of such a pylon is that it should withstand the combination of a horizontal load, due to the pull on the cable, and a vertical one due to the weight; both without exceeding specified deflection limits.

Well, the proof of the pudding is in its eating, you say, so let's try it out. Full size; then there cannot be any argument. So you sit down to work out in detail how to organise it all. The first thing that may occur to you is that the pylon will have to be mounted on a totally rigid and therefore a very expensive concrete base. You want to learn how much the structure deflects, and not its foundations. And you will need some kind of crane with a carefully calibrated loading device hung up on it, plus all kinds of cables and pulleys. And how about measuring the deflections – do you use surveying techniques, with laser beams, or what? And suppose the deflection in one or other direction is too large (which means the pylon may collapse) or too small (which means you are wasting money on unnecessary strength) what part of the design is best modified to correct the structure in the most economical way, will new foundations be necessary, and how many attempts shall I need before all this interlocking complication is sorted out? And will I reach retiring age before it's all completed?

Faced with this daunting array of delays and complications of a direct test, you consider an indirect one. In a totally direct test everything is represented by itself, in a totally indirect one everything is represented by something else; in this case, probably electric pulses in a digital computer. Programme the mathematical analysis of the behaviour of your structure and that of your oval steel tubes, and you can read off immediately, to any degree of accuracy you want, all the deflection figures. And you can instantly find and compare the effect of using different thickness of tubes in various places. It is ideal for the job; but has it been too idealised? Do unorthodox shaped tubes really obey the textbook formulae? And will the welded joints modify the axial and shear flexibilities that you have assumed? Sometimes a paradise may be a fool's one.

It would be exceedingly comforting if you could devise an indirect test that would allow the behaviour of steel in an oval section to be represented by itself, for then there could be no argument about it. Well, one can devise a combination of direct and indirect tests and it is a most useful technique.

It is usual to make a model. Until fairly recently a model either meant a small toy or a young lady. Now it also has a specialised research meaning, usually based on the theory of dimensional analyses. This theory allows you to scale down your pylon to a handy size; a model that you could build and test in a laboratory. It would have little steel tubes and welded joints; you could hang instruments all over and apply forces in useful directions and magnitudes. It would be an entirely indirect test, except for one thing. Everything would be represented by something smaller than itself, but the steel would still be directly represented by steel.

If you were interested in the deflection caused by the wind, you would put your model in a wind-tunnel. Then everything would be indirect except the steel and the air. For this test, you could replace the steel by suitably shaped wooden slats and then only the air would be self-representative.

In the last chapter we referred to 'ducks' for exploiting wave energy and how compliance in the mounting, which you or I might never have imagined worth investigating, added markedly to its efficiency. This was established in a model test where everything was represented by 1/100 something else, except water, which was self-representative (see Figs 3.1 and 3.2).

Salter writes about his experiments as follows.

'*Moving Axis Results*

'Several authors, Count, Glendenning, Mei, and Standing have published theoretical efficiency curves for ducks and suggest that movement of the mounting reduces the performance.

'It is true that the early ducks were intended to operate on a relatively stable reference frame and movement was regarded as necessary for survival. But it is not true to suppose that there is a continuous degradation of per-

Fig. 3.1 A 1/100th scale model duck in calm water.

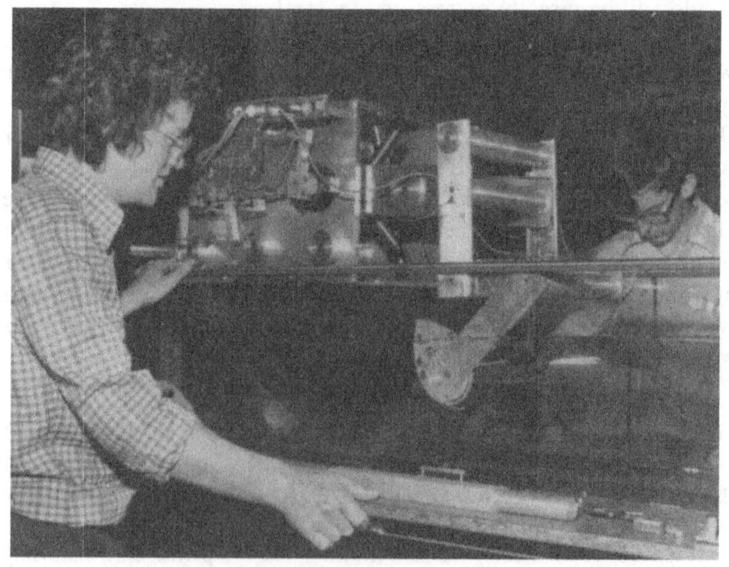

Fig. 3.2 A time exposure of the same model with tracing fluid showing orbital motion.

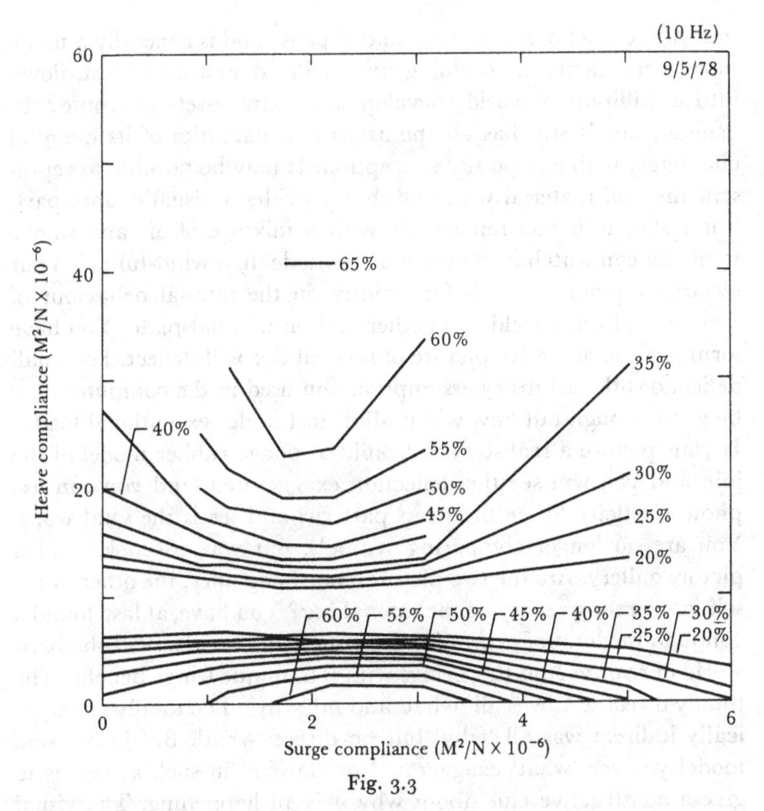

Fig. 3·3

formance from absolute rigidity to total freedom. The experimental evidence is more complicated, more interesting and more promising that the straight degradation theory would suggest.

'Our mounting allows separate control of the rigidity of the model in heave and surge. We made efficiency measurements at a number of frequencies and a number of combinations of rigidity. A typical result is shown in Fig. 3·3.

'The horizontal axis is compliance in the surge direction and the vertical is compliance in the heave direction. The wavelength corresponds to 22 duck diameters and the curves show contours of efficiency. There are two separate regions of good performance and neither occurs at the point of absolute rigidity. They are separated by a region of extraordinarily low efficiency. We call this area "death valley". It is a feature of tests at all frequencies, and may provide a useful operating condition for survival.'

A model test is a direct test looked at through one end of a telescope or the other, with or without a material substitution. Our immediate purpose is to decide on the potential efficiency of this technique when applied to a new idea in principle. It has the immediate asset of saving time and expense and is generally a much more convenient way of doing things. But does a direct test, flown into a Lilliputian world, develop any extra assets en route? In general, no. It still has all the assets and liabilities of its essential character; with one possible exception. It may be possible to represent the real material with one that provides a visual counterpart. For instance, if you replace air with a mixture of air and smoke then you can watch its flow around a model in a wind-tunnel. Your electricity pylon depends for rigidity on the mutual behaviour of several oval tubes welded together and on to a flat plate. You have formed an imaginative picture of how all this will deflect. For small deflections the arbitrary assumption you used in the computer may be good enough but how will it all distort under exceptional loads? Is your picture a realistic one? Build a sponge rubber model of the join and you will see the deflection exaggerated, and you can use photo-elasticity to see the stress pattern; and see is the vital word. You are no longer comparing symbols, but you are looking at a picture gallery. Are the two pictures, one imaginary, the other in the self-interpreting reality, of the same thing? You have, at last, found a common medium of expression for your intuitive idea and the hard world of reality. And this brings with it a unique fringe benefit. The totally direct tests was all 'what' and no 'why'. The totally electronically indirect was all 'why' but no direct 'what'. But in a visual model you see 'what' exaggerated or clarified in such a way as to give a constructive clue about why it is all happening. The visual model is the only research technique that can give an objective 'what' and 'why' simultaneously in trying out new ideas in principle.

Looking back over all these combinations of direct or indirect research I think it is necessary to repeat again what was initially emphasised. In all indirect tests, or the indirect parts of a combined type, accuracy is totally dependent on the reliability of the representation and the re-interpretation. If your representation has a possible error of $\pm$ 10%, and your interpretation the same, results that should appear identical may vary by nearly 50%. This source of error may be easily overlooked as it is, so to speak, outside the visible area of the experiment. You can make huge abstract errors without anyone falling over them in the laboratory. They are modest to the point of invisibility. One of my most vivid experiences of this occurred when I was consulted as an independent party over a highly inflammatory controversy. Again I must be careful to make the parties concerned unidentifiable. But this is not a fairy story, far from it.

Maintaining therefore a tactful ambiguity, I can state that a certain company invented a completely new way of carrying out a certain chemical change. It was exploited in a very large and complicated installation which processed an expensive liquid in huge containers. The company, not being manufacturers but engineers, received orders to modify the existing equipment and it was agreed by each of three different factories to pay them, as patentees, a large sum per day for the use of their invention. Everyone was as pleased as Punch. The engineering company then approached a government department for certain financial and export assistance and this department in turn sent down a research team to confirm the efficiency of the new process. Their report, in due time, stated that, in their view, the new process was no better than the old; might even be a bit worse in fact. There immediately followed the most hair-raising row it has ever been my embarrassment to behold. I am still convinced that all parties were being strictly honest at heart, but this was not a universally accepted view at the time. I was asked to go down to one of the factories and so organise a research programme that I could submit a clear and query-proof verdict that all parties could accept. The situation was tricky, the atmosphere electric, the money involved large and the time short. I decided on a simple, partially indirect, test. A series of small samples, representing the liquid before processing, were to be taken; the full-scale process carried out and a further series of samples then taken and the results compared and interpreted. The testing of these samples was largely a chemical process. I am not a chemist. I was therefore in the hands of the local experts; I did not doubt their honesty but I wondered about their chemistry. I went down to the factory, was given an office to use, and a flock of white-coated gentlemen to command. I decided to start by taking twenty samples representing the liquid before processing and twenty after, out of both the new and the old-type production lines. I personally supervised the taking of each sample, carefully labelling them; every move being followed and every label copied into notebooks by my white-coated attendants. I could not help thinking of Snow White and the seven dwarfs.

At the end of the afternoon my office was filled with trolleys of bottled samples for the chemists to test next day. Everyone thought I was testing the process; actually I was planning to test the chemists. I said I would be working late and they need not stop behind for me. After they had gone, I changed all the labels around on the bottles, carefully noting which was now on which. Next day everyone was in bright and early and the trolleys of samples were wheeled away into the chemical testing department. I asked for the results to be ready as soon as possible and sat down to wait. By coffee time nothing had happened and so I strolled into the laboratory and asked brightly how they were getting on. They apparently weren't. Just before lunch the chief chemist said that there must have been some muddle over the labelling; it was essential to know which samples were taken from the old process and which from the new. I sat back in my chair and said 'why?' And so the answer had to come out. The limits of accuracy in their chemical testing meant that the results could be interpreted quite widely. The chemists were convinced, on other grounds, that the new process was better but they needed my notes on the new labelling to arrange their results to show this.

What happened next is almost anyone's guess; I'm not saying.

4

The basic idea—competitive research

This method of research provides unrestricted efficiency over a restricted area of application. The essence of all competitions is that they should be fair, with no bias in favour of a particular entrant. A game of tennis is basically fair, for the players hit over a common net, all balls are identical and any effects of the weather cancel out because the players change ends periodically. The words 'common', 'identical' and 'cancel' are the vital ones in a tennis tournament and also in the competitive method of research. Its great asset is the self-evident nature of its results. It is the shoppers' technique. 'Let's choose this drawing-room suite, it's more comfortable than the other one.' 'Your new train suspension,' says the small boy, 'makes me feel even worse than the old one.'

This win-or-lose verdict is only a valid one if it is largely or entirely an objective one. The little boy's judgement may have been biased because he had just golloped three ice-lollies. The sofa seems the most comfortable because you are worn out from shopping and the softer and sloppier the springing the better you feel at the moment. And so we must add 'non-subjective' to the words 'common', 'identical' and 'cancel'. The usefulness of the technique lies in its problem.

Asked to advise on the design of concrete motorways and air-field runways suitable for high local loads, I was appalled at the number of variables that were all clamouring for a place in my research. The concrete could be made of different strengths and thicknesses, with or without steel reinforcement. The size, shape and location of such reinforcing steel could be in dozens of varieties, so could the pattern and magnitude of the applied loads. Direct full-scale tests of slabs could, in theory, be tried, but in practice what was the criterion of potential disintegration, how could it be instrumented, and how far would the load pattern affect it all? The primary purpose of the concrete was to bridge gaps or subsidence in the subsoil so there was not only a pattern of load above but a pattern of non-support below. How could one sort out this bag of tricks? The answer lay in a competitive research technique. Cast two concrete slabs of identical rectangular shape with the types of reinforcement and concrete thicknesses you wish to compare. Stand them on edge parallel to each other and hold them together with large-diameter bolts going right through each other. Keeping the slabs apart is a pattern of rectangular blocks. The position of the bolt-holes and the blocks can be chosen at random or systematically. The nuts of the bolts are then progressively tightened so tending to distort each slab against the other with an identical force as the bolts pull in and the blocks push out. The slab shapes are identical, the bending moments may be incalculable but will be identical and the weakening effect of the bolt holes will cancel out. As you tighten the bolts (or increase the tension hydraulically in thicker slabs), you walk round and observe the only variable; i.e. the cracking and approaching collapse of the slabs. It is a straight competition between the two. Being a direct test it will tell you nothing of why it all happens but you have been able to observe, quite realistically, what happens. In this way, if you wish it, you might be able to evolve a design by a series of competitive tests without any theoretical background but by the use of highly objective results. It would take time but might be the only way of doing it. A more sophisticated example of the same basic method was used to determine which type of pump should be selected for the Tehachapi crossing scheme in California. Here a man-made river, the size of the Thames, had to be pumped over a 2000 ft-high mountain pass to cultivate an area of arid, but otherwise perfect, climate. Three firms submitted three differently designed pumps. The money involved was immense and the best design had to be selected uncontroversially.

A most complicated high-power test rig was built at the National Engineering Laboratory in Scotland and model pumps were tested against the full 2000 ft head, using powers up to 4000 hp. All the bearing arrangements were the same for each pump and they used common conditions in every way. There could be no argument about the verdict; it was a fair competition; the best man had to win. But no-one except himself knew why. The manufacturer could keep his secret still. 'What' was self-evident, 'why' non-evident. The competitive test is always limited in scope because you cannot play two games at once. You might well say that the behaviour of those concrete slabs would have been better compared under conditions of common deflection rather than common vertical load. And you can easily arrange a competition on these lines too. I did.

Clamp down a pair of rival slabs on to a rigid foundation with a pattern of blocks of uneven thicknesses under each. Again compare what you see. But you cannot play squash and tennis at the same time, nor compare both common load and common deflection in the same experiment. You can answer only one question at a time. Now this looks so useful and straightforward that it is tempting to adopt this research technique on every possible occasion, even where it is unwise to do so. Essentially the technique consists of making constants identical, and the relevant variables either common or cancelling. But as you do not know for certain either before or after the experiment why it is all happening, how can you be certain why the variables you have chosen are, in fact, the relevant ones? The competition may be basically unfair. Someone's boxing glove may have a brick in it.

As has already been mentioned I worked in an engineering factory during my university vacations; a necessary qualification for my degree. The firm was largely concerned in manufacturing machinery for making paper. An immediate problem on their hands was the design of a centrifugal pump for circulating paper pulp round a closed circuit. The pump was required to handle a large quantity at a very low head. And the liquid was a mixture of water with up to 5% of solids; rather like sloppy porridge. I offered to design a pump, and although my practical knowledge of pump technique was nil I guessed away happily at what I did not know. The possibility of cavitation (although I did not know to call it that) was an obvious danger, and a vast amount of power could also be consumed unless the head fell off rapidly as volume rate increased. I had a lot of fun designing (or mostly guessing) inlet configuration and blade angles. I gave them a full detailed drawing of my design and departed to Cambridge. Next vacation I reappeared and asked after my pump design. 'It didn't work,' I was told. 'We tested it against our standard design; yours would not pump anything, even if we ran it backwards, but our design shot the water right up to a high tank.'

Pink with embarrassment and seething with disbelief, I was taken to where a test rig had been set up. Each pump in turn had to transfer water from a ground level tank to one some 30 ft up on the factory roof, through a small bore pipe. I pointed out that my design was for pulp, not water. They said that if it could not pump water it certainly couldn't pump pulp. I next said that the requirements were for a self-throttling circulating pump giving a few feet of head. They said that their rig was designed to make sure there was plenty of performance to spare. They pointed out that in a perfectly fair test their pump had won; mine was a flop. They were all 'practical people' and made little attempt to listen to my protests that the test conditions were actively misleading. I soon gave up. As none of them knew, in theory, why a centrifugal pump worked, none of them knew why the tests were completely misleading. You cannot often totally isolate 'what' from 'why'.

Thus the competitive method of assessing a new idea in principle has the advantage that there is only one main pitfall and the disadvantage that this is a huge one. So beware. There is, however, one certain safeguard. With competitive research, within the context of the direct, or partially indirect, technique, if you can define in experimental hardware exactly what you want, you can exactly compare results. 'Exactly what' are the vital words. Totally pragmatic, with no theoretical conjecture, these magic words allow you to proceed with confidence, with full hands and a studiously blank mind. The precise function of the California pumps was known; no-one could agree on what was exactly wanted with the paper-pulp one.

With the competitive we have completed the list of the most widely used research methods for assessing new ideas in principle. There are a number of others but to include them would make the whole subject unwieldly. It is no good being exhaustive if it means being exhausting, and I think we are now in the position to summarise what we have been saying about initial development.

Efficient research can make self-evident the workability and sometimes the validity of a new principle but the process is irreversible. Concrete experiments will never, unaided, arrange themselves into an abstract pattern. They can confirm theories but not initiate them. Chessmen never play chess.

Efficient development hinges on the choice of the most suitable technique.

We might usefully summarise the assets of the various possibilities we have discussed in the form of a table (Table 4.1) giving each marks for merit.

Our subject is the new idea; our purpose to find out about it in principle, subdivided into four different objectives – what happens, why it happens, the possibilities of new materials and the potential of energy sources. The main research techniques, as we have seen, are the totally direct, the totally indirect, a combination of them, the visual model and the competitive. But for the sake of realism in our chart, we will somewhat simplify these titles and call the totally

Table 4.1.

The new idea in principle	Full scale	Electronic	Model	Visual model	Com-petitive
What happens	***	–	*	*	****
Why it happens	*	***	**	****	–
Energy source	–	*	****	***	–
Material	****	–	*	–	*

direct 'full-scale', the totally indirect 'electronic', the combination 'a model', and leave the visual model and competitive description as they are. Marks will be awarded on their potential research efficiency, with four stars for the best, with one star for the least promising and nothing for the probably unsuitable.

5

First embodiment

The castle in the air is generalised and abstract. In total contrast, the castle in the market place is a particular castle, made from concrete materials. You would fall through the floor of an abstract castle. The first embodiment, the first castle on the ground, is composite in nature. The general ideas come down to earth in it, a vertical descent into reality. But it must have its windows open towards the market place. While being true to its vertical fall-out of ideas, it must incorporate those practical and commercially orientated features, fed across horizontally, that will make it saleable. The first embodiment stands on that common ground where the particular and general meet. The horizontal and vertical intersect in a distinctive pattern. And the danger is that the common ground may become a battle ground. There are two ways in which such a conflict becomes almost inevitable. The first is to produce and test a prototype that has its head totally in the sky, its design chosen and slanted to make the general idea its most spectacular and clever-looking. 'Hang the expense but knock them for six' is its motto. The other disaster is to collect a mass of reputed essential practical points, some of them possibly self-contradictory, and insist on all of them being inflexibly and simultaneously incorporated. One is an invention no one wants and the other wants something that no one can invent. The first essential step to prevent these and other types of civil war – sometimes not very civil – is to make every effort to ensure that the requirements of the market place are communicated as far as possible in figures and not in words. Both these requirements and the responsive ability of the prototype to fulfil them should be capable of being independently audited.

When Whittle designed his first embodiment of a jet engine, as we have already remarked, he was given a minimum objective of 500 lb thrust. This could be independently audited by watching the numerals on a dial. The dial was not concerned whether the invention was worked by Whittle or witchcraft or a combination of both.

The numerical value of the actual thrust was the only thing with which it was concerned. Numbers speak louder than words, and in this context are a great deal less open to misunderstanding. But it is equally important that the shopping list brought back from the market place should (as we have already stipulated), have its value in numerals too. Words such as 'cheap', 'light', 'quiet' are hot-houses of subjectivity and misunderstandings.* And the first embodiment must avoid avoidable misunderstandings – there may well be enough unavoidable ones. Writing down figures clears the air and often clears people's minds too.

All prototypes are in the numbers game in a rather more sophisticated way as well. It is essential that their internal behaviour is monitored numerically. With the idea in principle, algebra may well be all we can expect; prototypes must learn arithmetic. Only when you put numerals in formulae do they become alive. You must use proper measuring devices.

The next essential is that the numerical measurements must be taken simultaneously, (repetitions do not always repeat; there are exceptions to this rule, but not many). You must have a realistic numerical picture; one that you can sit back and look at afterwards. Lastly and most important of all is that input variables must be numerically controlled and recorded, and if they are interrelated feed them through a computer to sort them out. Ignore some or any of these rules and you will be in trouble.

In my first year as a student in Cambridge, I was spending most of my time designing a sprint racing car and collecting bits and pieces to be used in or modified for its construction, which was highly unorthodox. Searching through the advertisement columns for suitable second-hand, and so cheap, components, I discovered that a racing carburettor was for sale. It was called a Bink's Mousetrap. I was quite unable to resist a name like that and wrote off for it. Anyway it was very cheap. Later I was to discover why. When it arrived I looked at it in awe, for it was made of solid, gleaming brass, with a long variable-venturi choke from which it obviously derived its name. No mere drawing could convey its social impact. It looked like an ancient blunderbuss and had apparently been designed by a madman. It had three jets for the petrol, queueing behind each other, a separate cable-operated control for the venturi, another for the butterfly throttle – five variables in all.

It was late one August evening when I at last completed the chassis of the car at the back of a factory in South London. I was

* 'The Design of Design', page 7.

determined to drive it to my home in Kingston and put the body-work on there. I had already entered it for its first race in a few weeks' time. I pushed the car through the factory gates and a curious crowd immediately began to assemble. I selected three jets, put them in, adjusted the choke at random and invited the crowd to push. This they did with enthusiasm but the engine made no attempt to start. I juggled the jets and tried again – no result. The crowd was becoming exhausted and night was beginning to fall. A grocer's van had stopped and its driver kindly offered to give me a tow. We found a sizeable rope, I juggled the jets again, putting some of them in upside down for luck, and we were gingerly towed away amid general cheering. I let the clutch in; nothing happened. I signalled to the driver to speed up, waited until he was doing about 40 mph and then let in the clutch in a low gear. The engine, I saw from the rev-counter, was doing over 3000 rpm. A fraction of a second later it started up with a noise I have never heard equalled. Admittedly it had two straight-through exhaust pipes – the two silencers that I had hammered on to their ends as a gesture towards legality had immediately flown off when the engine started and had departed into the gloaming at high speed. We stopped, While the terrified van owner hurriedly unfastened the rope I kept the engine running, which meant keeping the rpm above 3000. Part of the incredible noise, I suspected, was because I had the valve timing badly wrong. The exhaust valves were opening during the firing stroke and the inlet valves partially during the exhaust stroke. This also accounted for the petrol that was being blown out of the air intake and was settling in pools around the Bink's Mouse-trap; next there was a shattering blow-back through it and the pools of petrol immediately caught fire. To put out the fire I had to speed up the engine even more and so suck the flames back into the carburettor. It was immediately apparent that the relationship between the throttle opening venturi aperture and engine speed was interlocked, sensitive and obscure. By this time it was dark, apart from the flames from the exhaust pipes, but I was determined to drive home, although I had no lamps, number plates or any other legal trimmings.

I drove the portable bonfire through the suburbs of Wimbledon, stopping to suck out fires periodically with a particularly deafening din and, finally, reached the Kingston by-pass. Half way down it I reached a battery of flashing lights and a police car. They had set up a road block to stop me. I came to a standstill, revving the engine hard, and the police gathered round gesticulating. I gathered that they wanted me to stop the engine so that they could talk. I turned off the ignition and in the sudden silence that resulted, I took the opportunity to remark that as they had stopped it, they must also be responsible for restarting the engine for me. Quite a long conversation ensued and my name and address were duly recorded. Finally the policeman in charge told me that I was quite impossible, and they all got back into their car, slamming the doors. But just before the last door slammed, one of their voices said quietly 'there is a rope in the back'. I could only interpret this in one way. I went round to the back of the police car, opened its boot and found the rope, which I carefully removed. Pushing my car to the rear of theirs, I connected them with the rope, got in and waited. After a moment the police car slowly drove away, all heads looking stonily ahead, officially unaware of an highly illegal vehicle they were illegally towing. A quarter of an hour later the police car gradually came to a standstill outside my home, all heads still staring ahead. I quickly untied the rope, put it back in their boot and remarked to the night air that our policemen are wonderful.

You will notice that the only numeral in this experiment was the number of policemen, and interconnected variables abounded. No wonder the results were not entirely edifying.

Recently a large organisation hit on a novel process system and carried out research into the idea in general. So pleased were they with the results that they omitted any research into a first embodiment ('no time to waste on that'), and spent a vast sum building a complete production line. After several months of production not a single saleable item had been produced. Everyone was depressed, especially those who had to pay for it all. Asked to advise, I entered the factory and the first thing I saw was a complicated machine with two operators controlling the process by altering the settings of three essential variables to compensate for variations in the raw materials. But altering one setting immediately influenced the effectiveness of the other two; although not obvious unless you were looking for it, the controls were interdependent. No-one who had had dealings with a Bink's Mousetrap could fail to appreciate the problem, however respectably dressed up. I knew instantly that the process could never be controlled manually; nor could a computer be used until the right numerals were found.

Economics demand that interlocked variables must be recognised and the problems that they offer solved at the first embodiment stage. At the final stage the cost of the factory-scale experiment may well be horrific.

As we shall see later, the most efficient research strategy will depend largely on the class of prototype, but there are some general points that usually apply to all of them.

The first is that while a new idea in principle is normally positive and exciting, the first embodiment inevitably appears negative and frustrating, often plagued by irritating delays.

If a prototype works perfectly first time and needs no modifications; and you knew this would be so in advance, the making of it would have been an unjustifiable waste of money. Prototypes justify their existence by not working first time. This middle third in the history of the development of a new idea into a marketable product is much the most difficult; it is the phase of failures. Often detail design features, not directly dependent on the new idea itself, make things grind to a halt, blow up or fall down. 'It will never work' mutter the onlookers; just loud enough to be heard. Prototypes need a clear head, a stout heart, and about three times as much money as you originally thought. And something even more vital too. A determination to embody the original research-supported idea and not to 'improve' upon it. Such 'improvements' are in general both compulsive and disastrous. That they should be so compulsive is, at first sight, rather odd, but there is a logical explanation for it.

A few days ago I heard an inventor say with relief 'now all they have to do is to make the prototype'. Note not only that the word 'all' is almost certainly going to be vastly complicated as distinct from its implied simplicity, but also the obvious delight of the inventor in handing it over to someone else. Now he can sit back and relax for a change. The months of concentration on the problem are over; he can take a decent holiday at last. And he does. But his subconscious does not. He has unwittingly produced, in a most concentrated form, that concentration–relaxation sequence that we have already recognised as the basis for inventive ideas to be subconsciously evolved and handed up into the conscious. The inventor has no defences against a string of modifications or improvements bursting into his imagination. In addition, he is, by this time, thoroughly restless and irked by the delay in drawing and making the prototype. The combined result is an irresistible urge to ring up the company doing the work. And our imaginary inventor will most probably have a long conversation, which, from his end, sounds something like this: 'Is that . . . , could you put me through to the chief draughtsman . . . , hullo John, how are you . . . , now look I've just had a marvellous idea about modifying the prototype . . . , this is what I want you to do . . . but I'm sure it won't take long to do the drawing again . . . Well, I hadn't really thought about that but I suppose stainless steel would do . . . I'll leave these details to be worked out on the drawing board . . . you won't have to alter the main framework we tested very much . . . Well, I don't think it should affect that . . . I am sure I can leave that to you . . .' and so on.*

There is only one thing to do with an inventor like that (and nearly all inventors do behave like that on occasions) and that is to tell him, kindly but firmly, to pipe down. If you don't you will, almost inevitably, be in for a load of trouble, for it is precisely these well-meant but hasty modifications, devoid of a research pedigree, that will bedevil the behaviour and befuddle the results of the first working embodiment. And the reason will be that it is not the first embodiment of the new idea. It is the attempted embodiment of an idea and a half. And the half has bypassed the discipline of the initial research. In practice, the division of research into the categories we have chosen is not just a convenient way of writing a book, or even that overall analysis so loved by the academicians, but a pattern and discipline of thought without which efficiency in research is most improbable. The first step in research into any idea, original or a reputed inspired modification, at a later date, must always be into the idea in principle. An idea without established principles lacks the necessary upbringing and will usually become a deliquent and a drop-out. The fact must be faced that the first embodiment of an idea will always be out-of-date, to some extent, with the inventor's latest thinking; the time interval makes this almost inevitable, but unless you freeze the design and get on with it, second best though it may be, you will regret it. Something good will be good enough, the best may be its enemy at the moment.

The second generalisation we can make about first embodiment is summarised in the word compatibility, and there are three areas where it applies. First there is internal compatibility.

The vast majority of inventions are not of totally new machines or structures but of a new element within existing ones. After all, in theory anyway, no device will be, in its embodiment, totally novel, for it must include old-hat components like nuts and bolts, transistors, or reduction gears and so on. And the new must be compatible with the existing. You must not put new wine into old bottles.

This was first forcefully brought home to me many years ago, in about 1936 I think. The famous car firm where W. O. Bentley

* 'The Science of Design', page 7.

produced his classic sports cars had gone into liquidation, and W. O. Bentley had joined the Lagonda Car Company. He designed a new model for them and a prototype was built. Basically it consisted of a $4\frac{1}{2}$ litre engine, put in a $3\frac{1}{2}$ litre chassis, with an aluminium two-seater body. It also had immense Girling brakes of a new design, fitted to a motor car for the first time. The project never went into production, the prototype was too fearsome, so they sold it to me. It was a delightful car but it certainly had some oddities that possibly made it unsuitable for the average owner. For instance, when driven fast its exhaust manifold became almost white-hot and often set fire to the ignition system. One night, at a party, I met a journalist who was interested in learning how the car performed. I suggested a test run. In the early hours of the morning, even in a London suburb, there was, in those days, almost no traffic and I urged the Lagonda along happily. Now in London it is very rare to discover a side road that abruptly incorporates a steep downhill hair-pin bend. However I discovered one that night when the Lagonda was travelling at a quite cheerful velocity. I stood on the Girling brakes. They were superb; a little too superb in fact. The sudden locking of the huge front wheels produced such a torque in the front axle that the front springs collapsed, and allowed the axle to batter up against the chassis members, thus making the car virtually unsteerable. The car was now travelling broadside on towards a substantial but low brick wall, which we hit at a very considerable speed. We knocked it completely flat without, as far as I could see, slowing down the Lagonda very much. Admittedly I could not see a great deal as the immense headlights were now pointing vertically into the sky. The car then bounded down a very steep slope covered with fruit trees, and cut them off short. Next we arrived at the rock garden, whose jagged edged stones tore away all the brake operating rods beneath the car. The foot brake pedal descended to the floor boards and I remarked to my passenger that in future we would have no brakes: we ran lengthwise over the grass tennis court without losing much speed and finally stopped amongst the rhododendrons where the garden sloped uphill again. I climbed out of the car; searched for the house in the dark; woke up the owner and explained my little difficulty to him. He was very nice about it; in fact I used to send him a box of chocolates on each anniversary, but never without remembering the marked incompatibility between the new Girling brakes and the ancient suspension system of the $3\frac{1}{2}$ litre chassis.

Pitifully expensive as it was to build, it is your duty and, indeed, in your interest, to try to make your prototype break down by every legitimate means. If it is a new section spliced into an old machine, watch especially carefully the frontier between the new and old. You will, of course, be tempted not to watch this at all, you will be concentrating on the behaviour of the machine as a whole. The ideal frontier is no frontier at all, and this is what you must always be aiming at. Mathematics knows no frontiers and it is by mathematics that a machine works. You cannot tell a differential equation to keep off the grass.

But often at the prototype stage an interloping frontier is unavoidable. For instance, imagine that you have invented a new heat exchanger and must try out a prototype version to see if it works and is compatible with the rest of the machine. Normally, let us assume, all the pipe joints would be welded and the whole machine would look a unit and behave homogeneously. But you must obviously make facilities for taking the heat exchanger in and out for adjustments or modifications, and so you will be using flanges and bolts instead of welds. Now these visible frontiers may alter the performance in an invisible way. Expansion effects, rigidity, vibration, inertia and so on may help or hinder the behaviour of the total machine. And don't be too hasty in saying 'that won't affect it' unless you know with confidence both the 'what' and the 'why'. And don't let wishful thinking produce the opposite error. 'I am sure it would have worked properly if the joints had been welded' must not be a final verdict. You must weld up the joints and see, expense or no expense. A prototype is there to tell you things, not vice-versa.

For instance it may tell you to regard the interloping new idea as the basis of the whole design and make the old conform to it. An example of this was the way that the synchromesh system was introduced to gearboxes. At first it was a 'goody' stuck on to the standard design. Now it is often the dictatorial basis of the total design.

The next generalisation concerning first embodiments is that they must be compatible with an economical working life. The inexperienced inventor, who quite often comes up with the brilliant ideas, nearly always overlooks this. Time still dictates.

In automobile engineering the useful life of a car is assumed to be about 100 000 miles. If driven continuously it would therefore last three months. A ship's engine must have a life of 15–20 years, running continuously. If a car was built to the same life scale as in this example of marine engineering, it should last 400 years. A prototype must embody an idea in the style of engineering that is compatible with its expected working life.

No one is asking you to spend years seeing if you can wear out a prototype through old age, you won't have the time anyway. But you must choose the size, shape and materials for the crucial components that established engineering technology tells you to be essential for the appropriate life span. You must not cheat. There is no difficulty in making a four seater saloon car do 100 miles per gallon petrol; but it may collapse in a thousand miles or less.

But there is also the criterion of external compatibility. It is obvious that an engineering device should be at least temporarily impervious to its environment. If the common fault of the inexperienced inventor is overlooking time factors, that of experienced ones, history suggests, is underestimating environmental effects. Air and water are great blessings, we could not live without them. But they are mixed blessings. Air contains and transports corrosive chemicals and unforeseen violence, sometimes unexpected by the bridge builders, or cooling tower designers, and startles aeronautic engineers and everyone else. We tend to underestimate water, but it is the worst of the natural contexts. Of the new prototypes submitted for test by the American Naval Department in connection with new submarines, 100% have proved partially disastrous and 40% totally so, entirely because of the effect of sea water on them.

But at the moment I am concerned with a less obvious and self-assertive type of compatibility or the lack of it. A first embodiment has a foot in two camps. One in the abstract idea, the other in the concrete hardware. One in the symbolic, the other to be grounded in the real. One in principle, the other in practice. What we require from efficient research into a principle is that its effective working must be objectively demonstrated. But in the realm of the real we require economic efficiency too. Effectiveness and efficiency must be simultaneously embodied. And this does not automatically happen. If you want to squash a wasp on a window you can use a sledge hammer. It works, most effectively. The energy efficiency, in terms of swinging the hammer and picking up the glass, is poor. Money talks, and sometimes it says 'it's not worth the cost'. A prototype must stand or fall on its ability to be compatible to both an abstract idea and a highly economy-orientated civilisation. And this introduces us to the point I wish to emphasise, one which, at first sight, appears paradoxical. A simple idea, simply embodied, will be effective, but not often efficient. In practice simplicity is too simple minded to be efficient. The prime condition for efficiency is complication. For instance, almost every prime mover ever invented was delightfully simple in its original embodiment, and was also, by today's standards, horribly inefficient. To increase their efficiency by a hundred per cent we have increased their complexity by thousands. And it has been worth it. A box girder bridge is a simple concept; to construct and erect it efficiently is most complicated. An industrialist, seeing the prototype of an elementary new idea for the first time, is generally appalled at its complexity. 'Must it be so involved?' he asks. The answer is 'No, provided it is inefficient'. And there is a reason for this, I think.

It lies in the intrinsic complication of nature itself, and so the more compatible we become to her the more complicated we too must become. Basically nature is incredibly dynamic, there is nothing simply static in the whole universe. Like Alice we must run to keep still. This is both shown and symbolised by the work done in constructing the mounting for the new electron microscope at the Cavendish Laboratory at Cambridge.

A microscope of this magnitude must be insulated against shocks of all types. If mounted on normal foundations, it would be detectably disturbed by the waves breaking on the west coast of Ireland. A large building would normally be regarded as a stationary object, but in this case definitely not. Instruments were not only placed in the buildings but in numerous other parts of Cambridge, and a 24-hour-day record made of the extent to which everything jumped up and down. From these data a suspension system was calculated, designed and tested. But its effect was marginal. More sensitive instruments were next used and it was revealed that not only Cambridge but everything within it was resonating and vibrating continuously with an amplitude of about 1 micron. At this amplitude damping from natural causes almost disappears, and it is too small for the ear to detect. Everything, everywhere, is singing its own music; the word static is not a state but a simplification, and too much such simplification will not conform with the nature of Nature.

It follows that the prototype should be interlocked backwards to the initial research as exactly as possible and sideways to tumultuous nature; if you try to insert 'clever' simplifications you will be in trouble. I have been. In designing the prototype of an automatic self-feeding machine tool I used an electro-hydraulic control system. In the feeding process there were two separate operations that had to take place almost simultaneously. Seeking to simplify the design,

I arranged to operate the two movements together. The net result was a waste of time and money in separating them up again. Nature is impervious to wishful thinking, but it is a great respecter of quantum mechanics. Thus working efficiency increases hand-in-hand with compatibility to the complications of the natural, but so does the capital cost.

Both in purpose and practice a first embodiment has so far been represented as totally objective, factual, pragmatic and unarguable. And so it is, or should be, in general, but there are three exceptions: three important subjects about which a prototype remains tantalisingly dumb.

The first, and probably the most important, is the labour cost of assembly. Especially in products ultimately to be mass-produced the time spent fiddling about with a prototype in a laboratory is quite irrelevant. In production it would be done in a different way, in a different order, with different machines.

Anyone who has attempted to feed and thread through a new exhaust pipe into a car, dodging the clutter of its suspension and similar obstacles as he lies on the garage floor, never wants to do it again. Getting a camel through the eye of a needle is simple by comparison. I once asked the production manager of a very large car firm how his operators succeeded in doing this piece of gymnastics in the required 38 seconds, which was the set time for all operations. He replied that they never even tried. They first fastened the exhaust pipe up in the air and then built the entire car around it.

If an inventor remarks to you, 'it wouldn't take long to put it all together', he may be quite sincere but possibly quite wrong. The prototype will not help you much in deciding. There is an even more acute instance of this that arises in the evolving of a new material. A semi-handmade sample may overjoy everyone with its promise, but presents not the smallest clue about its overall production costs. On many occasions I have had to point out to an enthusiastic inventor, brandishing an impressive sample and a cost list of his raw materials, that if he could persuade a charitable institution to give him all his constituent materials free of charge, it would have only a marginal effect on the ultimate production cost. Power, computerised controls, cooling or heating chambers and a host of other expensive capital equipment, with their initial outlay and continuing depreciation, may often quite dominate the production cost figure. And the prototype is blissfully unaware of it all.

Fortunately most new inventions are not totally new, they are improvements attained by modifying old ones. Thus, much of the old costing structure may still apply. The difficulty is that the more radical the idea, the less the first embodiment of it will include old techniques. The more you want a thing, the less you may know how expensive it will be.

The final way in which a prototype can be irrelevant or even misleading, is on the fortunately rare instance where the new idea consists solely of a new appearance; underneath it's the same old idea, possibly a fancy looking tea pot, motor car or lamp-post. If your prototype is full size, well and good, but scale it up or down and you are in trouble. Small always (or nearly always) looks beautiful. A ship in a bottle looks nicer than a ship on the sea, even though one is in scale with the other. There is no equivalent dimensional analysis that is artistically valid. This is an unfortunate and expensive fact. Manufacturers of motor cars have no option but to build their body prototypes full size to judge their appearance. When the first Morris Minor 1000 was almost into production, it was thought that increasing the width of the body by 2 inches would be an improvement. So, an entire car was cut in half down the middle and a 2 inch strip was welded in. It was the only way to find out. The new appearance commended itself but some of the new bumpers had already gone into production and so they had to be modified. You can still notice early examples of this car with 2 inch spacers clearly seen welded into the bumpers.

The ear may still recognise the beauty of a melody after its transposition into a new key and different volume levels, but the eye finds size an inherent and characteristic element in judging the attractiveness of any object.

Drag it across a Lilliputian frontier and some of its magic will be added or subtracted by the customs.

6

Future compatibility

We have already discussed the necessity for the prototype to be compatible with both the inventive idea and the economics of the market place. But this is not the whole story: the prototype must be compatible with the appropriate technology as well. We have seen how, in most instances, a new invention tends to be ahead of the technology required to exploit it efficiently. The technology has to be invented after the invention.

But there are exceptions, which occur when the invention is, in itself, a new technology. Then, instead of an invention searching for a technology to exploit it, we have a technology sitting around waiting for an invention to pal up with it. We are in the midst of such an era now; the era of the silicon chip and the micro-processor. It is absolutely essential that a prototype must either incorporate or be consciously devised to be compatible with this revolutionary technique. It will change almost everything.

Odd as it may seem, inventors tend to move backwards in their prototype technology rather than forwards. They are possibly out of date in their knowledge of the subject, or else have a subjective feeling that it is 'playing safe' to use some traditional device. This is a fallacy for a drawing office should never be an antique shop. But this tendency is not solely a modern failing; it has happened all down the history of engineering. When I left Cambridge I decided to take a job as a manual labourer for six months to discover at first hand what a factory was like. I started work at 6.00 a.m. six days a week and my normal job was to be a general dogsbody around a vast machine, which was producing paper and incorporated its own prime mover, an ominous-looking steam engine. I walked round it in awe. To my untutored eyes the design seemed to be fairly modern, with one notable exception. The governor, which controlled both the speed of the engine and the paper machine directly coupled to it, looked as if it had come out of the ark, or at least a museum: it was a collection of Victorian brass-ware, driven by a doubtful belt. The foreman told me that sometimes the machine 'ran away'. Not clear what this meant, I asked what I should do if it did. 'You run too,' I was told.

Nothing untoward happened for several days and then suddenly the monotonous beat of the machinery began to rise to an ever-increasing screeching roar. The paper broke and the ground shook. I looked round for my fellow operators – they were already half way to the door. A few seconds later and we were all through it. All except one. The luckless foreman ran up the side of the machine towards the pounding engine end, stopped there for an instant to snatch up a paper bag and a large wooden spoon, climbed on to the framework and attempted to put spoonfuls of white powder from the paper bag on to the wildly flapping narrow belt that drove the governor. I felt that we were in immediate danger of losing the spoon; possibly the foreman as well. After some tense moments the belt began to bite, the machine slowed and we reluctantly returned. 'Playing safe' was not.

You must not shackle your prototype to yesterday's technology unless you are very much aware of what you are doing. Even today's technology will almost certainly be yesterday's by the time the final embodiment of the idea arrives. Only the future will become the present, and only prototypes with built-in compatibility with foreseeable technologies will not have to be extensively modified or started again. You don't have to install expensive technologies in prototypes if time or money is tight, but you must fashion within them an interface that will be broadly receptive to the technologies of the future.

But how can we guess the future? Some people are better than others at this, and one of the best non-technical books of its type is 'The Mighty Micro' by Christopher Evans (Gollancz). This deals with micro-processors whose technology will infiltrate and revolutionise almost all others with increasing speed. For the more specialised reader there is 'Micro Circuit Engineering' by H. Ahmed and W. C. Nixon (Cambridge University Press).

There is often another problem involving foresight which is unavoidable, difficult and nerve-racking. The cause is that many basic inventions need other inventions before they can be exploited. In some inventions no such mini-inventions are needed. Invent a speedometer for a car, a heater element for an electric fire or a referee's whistle, and you will not, with a bit of thought or luck, have to redesign the car, the fire, or the referee, I wonder if those who first thought up the idea of a nuclear power plant anticipated how many attendant mini-inventions would have to be thought up too?

Most inventions cannot be fitted straight into their existing engineering environment like a cork into a bottle. The development of a design needs the development of a host of little designs too, some of which may not be that little when you get down to them. Now if you have plenty of staff and plenty of money, everything can progress in parallel; any laggards receiving an extra injection of staff or money. But I suspect that this ideal state of affairs generally remains an inventor's dream. When he wakes up he is faced with a nightmare. His money is strictly limited, so is his staff. He cannot do everything at once; he must select which mini-invention to tackle first. It is a formidable problem. Rare it must be for any successful inventor of a complicated project to be able to look back in after-years and feel a comfortable glow because all his priorities had proved right. Far more often he would look back on maddeningly frustrating delays when the whole project stalled while some mini-invention was sorted out, or else the research was skipped, and it let down the whole edifice by failing at the critical moment, causing raised eyebrows amongst the financial backers and wildly improvised excuses from the inventor.

It was during the period that I chose to be a manual labourer in that paper mill with its unpredictable steam engine, that a brand new paper machine was erected in an adjoining factory and naturally I disappeared from my allotted task to see what was going on. It was, for those days, an impressive design, with variable speed motors for each section, ganged together by a crude but effective overall control. It was said to be the first of its kind in the country. Now, making paper properly is far from easy. The pulp arrives at the machine looking like milky water. To form the paper this milky liquid must be spread in an even thickness, leaving the fibres to form a thin layer. Subsequent processes of pressing and drying this thin and fragile film result miraculously in white, smooth and quite tough paper. But the most critical point of all lies in accurately metering the milky water across the draining mesh, from which it must soon be lifted off in a highly delicate state. A weak place anywhere, or uneven thickness, and the film will break somewhere, either immediately or in the subsequent operations: and paper, in sodden clinging chaos, will emulate Vesuvius in full spate.

If you upset a bottle of milk the contents will quickly spread themselves into a puddle of more or less even thickness. At first sight metering the paper should be equally simple. But in practice it is a major problem; probably the most difficult in the whole process. At any rate they were finding it so with this new machine; they had spent weeks trying to sort out the metering arrangements. When I looked at the machine it was, at last, in partial production, but only thanks to the ingenuity of the operators. As string contracts when wet and can be quickly attached to things, they had used it to tie on many improvised aids, such as bits of wood or canvas. This greatly helped the metering but did not do much to improve the appearance of the machine.

The annual dinner of the Paper Makers' Association was held a few days later and I had the effrontery to write for a ticket. Having done my daily stint as a manual labourer, I rapidly changed into evening dress and motored up to the city hotel where the dinner was held. I found that the Managing Director of my firm was arranged to be sitting next to the Managing Director of a rival one, and I was placed across the table in front of them. Soon after the soup my Managing Director started discoursing about his new machine with its superior electric drive. The eulogy continued during the fish course and was in danger of spoiling our meat course too, but at that moment my Managing Director stopped for breath and a long drink. I had a feeling that I had been left out of the conversation long enough so I remarked 'It is certainly a most unusual machine; 122 vital parts are held on by string; I've counted them'. This remark went over in a big way, especially with the rival Managing Director. My own boss looked as if he dearly longed to sack me on the spot, but feared that his rival would instantly offer me a job out of gratitude.

Don't, I implore you, listen to anyone who says 'We can easily sort out that little problem in the final machine'.

Is there any fool-proof way of assessing in advance what order of priorities should be given to the development of mini-inventions? I'm fairly certain that there is none. However, frequently faced with this dilemma, I have had, perforce, to evolve a personal system, but I present it with considerable hesitation. Don't, I implore you, accept it blindly; let it be a guide and not a dictator.

The idea is this. For each mini-invention you give values from two different scales of marks, and then add them up. Those with the lowest total should have the highest development priority.

The first scale of marks should represent the certainty–uncertainty ratio of the research into the defined design being successful. Suppose you allocate a possible highest mark of 10. If you considered that there was a 50–50 chance of the mini-invention being successfully developed as it stood, you would give it 5 marks. If you thought it far from certain, you might give it anything from 1 to 3, according to the circumstances. If you were very confident of the result, possibly only doing the tests to convince outsiders, you would allocate 9, or even 10.

The expectations and therefore the mark for a perpetual motion machine would be nil! You may have to assess the mark in a second-hand way; by deciding a figure via the advice of a specialised designer in a field in which you are not particularly expert. His confidence in his design may be influenced by the possibility of your employing him to develop it. This sounds most uncharitable but the possibility should not be ignored. In any case, inventors, not infrequently, fall in love with their own inventions. You may not be an exception. This is actually a good thing if it gives you that relentless determination to overcome all obstacles which has been the hall-mark of all the great inventors, but a bad one if your enthusiasm biases your judgement and you smother yourself and others under a suffocating heap of garrulous optimism. Self criticism is often both difficult and essential.

The second scale of marking is devised to represent how many back-up designs there are. How many other reasonable possibilities exist if your first choice proves a failure. If this is the only possible design you can think of, give it zero marks. It rates a high priority in finding an immediate research verdict on its viability.

If there are a number of other mini-invention designs up your sleeve and the research is largely to find the most suitable of them, give it 7 or 8.

Now add up the two sets of marks. You can see how the system is supposed to work by looking at extremes. A shaky mini-invention design, with no back up alternative would score say, 2 and zero; a total of 2. On the other hand, a design in which you have plenty of confidence, and when in any case, there are hosts of promising rivals, would rate 9 and 9, a total of 18.

Mostly, of course, the totals will lie well between these two extremes, and the development priorities will be indicated thereby. Tackle the low figures first. This method will certainly not be infallible in all instances but at least it will force upon you an element of disciplined thinking. In the next chapter we will see how this system can be applied to some contemporary problems.

7

The gamble

So far in this book we have assumed the existence of a main invention, with or without attendant mini-inventions, and have discussed how to develop it. But life is not always as simple as that; perhaps you would prefer me to say it is even more complicated than that. The main complication arises when you are faced with not one but a variety of proposed solutions, a crowd (sometimes two is a crowd) of rival inventions, each clamouring for your patronage. Unless you are a Government anxious to produce an atom bomb in a hurry you are unlikely to have the money to back all the options simultaneously. How then do you select which to choose, which to gamble your money upon? Before considering a specific example of this kind of choice we must clarify the basis on which it is to be made. In round terms, and as already discussed, the selection must be made on a personal assessment of the internal and external compatibility of the main inventions and the potential difficulties of their attendant mini-inventions. Added to these should be a highly subjective but often decisive verdict based on a sense of engineering style; the artistry of design. I have written about this elsewhere (*The Design of Design*, p. 20) as follows 'The sense of the artistry of engineering is invaluable but cannot be formally stated. A machine may look artistic in the normal meaning of that word without being good engineering. A bridge may look nice but fall down. An efficient high voltage insulator is often ornamental to look at, but if you designed it merely for perfection of form it would not necessarily be a good insulator. The artistry of engineering is essentially a matter of style and this is always a problem to put into words. We find the same difficulty in talking about style in music, literature or art. The definition is often only appreciated after the style has been recognised'. Writing about the world of science, Sir Arthur Eddington observed, 'We sometimes have convictions which we cherish but cannot justify; we are influenced by some innate sense of the fitness of things'. Perhaps this 'sense of fitness of things' is the nearest we can get to a positive definition of engineering artistry too.

Surely a self-aligning roller bearing commends itself as good in principle! And is there not a simplicity of style in the design of a squirrel cage motor? I sometimes wonder if such a humble thing as an umbrella is not a remarkable example of structural engineering. Very few structures can be erected or pulled down so quickly.

We must be careful to distinguish between a good principle of working and good workmanship in applying a principle. Probably most of us will remember the first time we prised off the back of a watch with a pocket knife. We were immediately struck by the fascinating delicacy and precision of the diminutive components, but this of itself said nothing about the essential style of the working principle. Did we go on past this perfection of workmanship and appreciate the economy of style in the principle? The precise metering of energy by a small spring from a bigger one is an engineering joy. The total energy that would only sustain a child's top for less than a minute, despite its low friction bearing, will run a watch for a day and, in total contrast to the top, at an exceedingly uniform angular velocity.

For transporting its own weight over rough ground, a child's hoop is artistically good while a motor bicycle and sidecar, even though designed to a high state of engineering sophistication, is clumsy in principle.

This distinction is important, for it is nearly always true that some new breakthrough in engineering design may initially appear less mechanically sophisticated than the highly developed traditional one it will soon replace. 'Style will always win in the end.'

I wonder if you have ever seen the Severn Suspension Bridge (Fig. 7.1). It is often regarded, rightly I think, as an artistic masterpiece. Delightful to look at, especially angled against a setting sun, it has been pictured in numerous paintings and photographs. See it if you can.

But when an engineer looks at it he may well see more than its beauty; or, if you like, another form of it. He will notice that the normally vertical wires that support the bridge deck are, in this case, at an angle to the vertical. And the reason for this only an engineer would know, understand and appreciate.

Today all large suspension bridges are primarily designed for stability rather than strength alone. Certain winds acting on certain designs could, and have, caused the whole thing to flutter and sometimes collapse. To guard against this has, in the past,

Fig. 7.1 The Severn Bridge.

meant considerable extra expense and weight. But the designers of the Severn Bridge came up with a most elegant solution. They established that by angling the vertical wires they could introduce built-in hysteresis losses that would damp out any oscillations. Economical, simple, a direct incarnation of physical and mathematical principles, it commends itself as self-evident engineering artistry.

Both the aesthetic appeal and the artistry of the bridge carry that same satisfying impact that is the hallmark of style in both.

We are now in a position to see how all this works out by considering the problem of selecting the most promising of various wave-

Fig 7.2

energy devices being put forward at the time I am writing. I am not attempting to tell other people their own business but merely using the problem as an illustration and an exercise, for it is not only topical but many of the problems are easily visualised or possibly remembered, if you, like me, are such a bad sailor you are almost sick in a bath!

I propose that we look at three different energy generating devices and I will use the data provided by a paper called 'Wave Energy', Energy Paper no. 42, Department of Energy, 1979.

The first is called a 'rectifier' and is shown in Fig. 7.2. Ignore for the moment the cut-away view in the right foreground and you will see what is, in effect, a house built on the sea bed with its top storey reaching out of the water. Along the water line are a series of what look like windows facing the waves, each about the size of a double garage door. Each is provided with flap valves at the entrance, those shown with vertical lines only allow the water in, and close up if it wants to come out; the other and alternating ones have flaps allowing water out only. The water entering fills a reservoir joining up the input channels, and another the output ones. If you now look at the cut-away section you will see that these upper and lower reservoirs are joined by a vertical connection which incorporates a water turbine.

The wave crests fill the upper reservoir, their troughs empty the lower, and the flow between generates electricity via the water turbine.

Well, what do you think of that? Would you gamble on being able to develop it successfully? First look at the main invention. No, we must start further back; first decide what is the main invention; the one upon which all the others depend; the centre of gravity, so to speak, of the design. The vital part is not necessarily the biggest part. Here, in my view, the design of the flexible gate valves is the decisive invention. Everything depends on them. They must be both internally and externally compatible with the conditions, and go on doing their job for many years. Maintenance would be an expensive nightmare. They must be constructed of something that combines flexibility and strength, and be non-corrosive too. We need to develop a specialised material.

The external conditions too are severe; pounding waves to be let through but not allowed back; an energy oriented problem of almost terrifying magnitude in a storm. In short, we must simultaneously develop a new material and new energy absorbing inventive design. And this is precisely what we cannot easily do.

If you look at Tables 1.1 and 4.1, you will see that the development strategies for materials and energy devices are almost entirely contradictory in every way. When it is essential to use models for one and essential not to use models for the other, how do you develop both in a single entity? I don't know. I am not saying that you cannot find a suitable material or design, I am not saying that you cannot find a number of them, what I am puzzled about is how do you know if you have found them?

Compared with this the development of the various mini-inventions seems comparatively straightforward, although you can never be quite certain. The water turbine has a low but not impossibly low head and, with luck, could be protected against marauding flotsam and jetsam. Perhaps it would be better not to regard the word 'luck' as an irreversible technical absolute. The reinforced concrete structure may well need the expertise that is being developed for oil rigs, but there is nothing particularly ominous in the proposed design.

Now let us look at the whole design as an electricity generating unit from the point of view of engineering style. One's first reaction is to notice that 95% of the project never moves; it just sits there. If you could find a suitable rock formation existing on the sea bed, exactly the shape you wanted, you would save most of the cost of the whole project. You aren't pouring money down the drain. Just into the sea. And as a continuous producer of electricity it is not particularly continuous. If the waves stop, so does the electricity, and quite suddenly too, with the braking effect of the stationary water on the turbine. The principle must work, the practice might work, but there is no feeling of artistry, no instinctively satisfying impact. It is hum-drum, with more hum than drum.

Fig. 7.3 Artist's impression of one form of raft converter.

The second device is called a raft converter and is shown in Fig. 7.3, which is almost self explanatory. Pontoons, floating on the water, are hinged together and the angular displacement at this joint is used to generate power. The main invention lies in this combination and presents some contrasting illustrations of development problems. The rafts are vertically compatible to the sea, they float, but incompatible in other directions. The waves would bash them all over the place if they were not strongly moored. Fortunately, as they can be constructed of known materials, such as steel and concrete, there is no self-contradiction in modelling techniques.

Of the mini-inventions, the outstanding is the power take-off system. The published account gives almost no details about this. In practice it would have to be split up again into two mini-systems. The first would be the hinges. Experiments plus direct mathematics would give initial values for the loads which must be immense.

The second mini-invention subdivision is the power-generating equipment. This starts off on the wrong foot. High power transmitted via slow angular velocity is an engineer's Everest. Torque makes things big; angular velocity you get almost free of charge. An air-driven dentist's drill exploits much more power than a grandfather clock but is not as large. Both the power take off mini-inventions have the developing awkwardness of massive size. The third 'mini' is the mooring chains or cables and anchors. This represents a contrasting collection of development problems. The anchors are up against (quite literally) nature's sometimes perverse and always vastly complicated and unpredictable variables. You cannot realistically imagine her in a computer or 'lilliput' her in a model; any more than you could a lady opera singer.

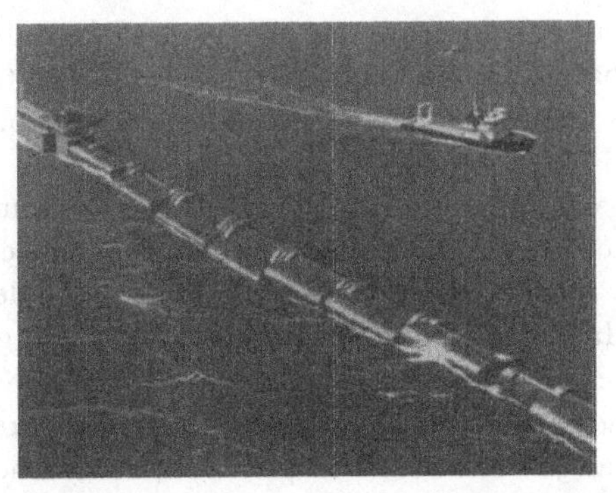

Fig. 7.4 Artist's impression of one form of the duck.

The same non-modelling applies to the savage sea water the chain has to endure. To develop anything from scratch would seem a daunting task. But chains and anchors have had to be used for centuries, and are a good example of evolved design. You will have to evolve them a bit further, possibly a very large bit.

From the point of view of style, the overall design has initially a familiar look. We know what simple floating rafts are like. But we must look again at them as an energy producing complex. They have much of the concealed complexity of the overtly simple. You can say it would be 'simpler' to build a house with one brick, but you would have to have a very large one and you would have to hollow it out in a most complicated manner. The 'complicated' design of lots of little bricks would be immensely simpler in every way. Provided complication really means duplication it is a great simplifier. A linked anchor chain has the complication of duplication and would not be economical nor work without it. Thus the overall style of raft is that of the mammoth; large, muscular (and extinct).

Finally we come to the 'duck' design which we have used earlier as a development illustration, Fig. 7.4 shows how a line of these ducks would look, each threaded, so to speak, on to a continuous and flexible spline, like beads on a necklace. The ducks operate with only their tips showing above the water and rotate about their spline as each wave goes by. If the waves are large and dangerous the ducks, because of their shape, automatically submerge and the resultant forces on them and their connecting spline tend to move the whole assembly forward, towards the waves, rather than having to be held by heavy chains and anchors from being swept away, as was the case with the raft. Moreover, in normal sea, the compliance of the spline not only increases the efficiency of the individual ducks but ensures that each in turn absorbs the waves, and the total force on them all is therefore again a fraction of what it would otherwise be.

Actually it would be possible to collect energy from the moving spline joints too. There is no direct mechanical connection between the angularly oscillating ducks and the power modules, which are built into each. These modules contain gyroscopes. A gyroscope is a heavy flywheel; if spinning around one axis, it is moved angularly around another at 90° to it, and tends to rotate or 'precess' around another axis at 90° to both. This physical phenomenon allows the power generating consortium to be totally isolated, working probably in a vacuum, with only electrical wiring passing through its protective casing. This arrangement would have other fringe benefits too. The gyroscope would provide an energy storage sink and could feed power back into the system when the waves were inadequate. The change over could be controlled within a few micro-seconds. And so a synchronous generator could be used; a considerable economic asset.

The power from the 'precession' of the gyroscopes would be collected by a large number of hydraulic pumps actuated by a large ring. They would feed a controlled rate hydraulic motor driving the generator.

Here the main invention must be the whole concept of ducks and their compliant spline. External compatibility is almost total. It is not a floating breakwater but a multi-directional absorber, self-tuneable to the variegated complexities of nature. The internal compatibility is equally good; with all moving parts working in a chosen environment where a technology of permanence could be evolved, and the whole working within a concrete corrosion insulating barrier.

Turning to the mini-inventions, the most vital are the hydraulic pumps. As we have seen, as energy producing entities they must be developed from small models. But in this instance we can use a short cut. Instead of saying to ourselves we must start with a model pump 1/200th full size, and then, when we have developed it, go on to gradually increasing sizes and rapidly increasing costs, we merely stick to our 1/200th scale design and use two hundred of them at once. They would have all the simplification of duplication, and you would have your eggs distributed within 200 baskets, not one.

The other mini-inventions are hardly inventions at all; flywheels, bearings and structures that use existing technologies or a develop-

Table 7.1.

Type	Main invention		Mini-inventions	Style	Total
	External	Internal			
Rectifier	3	3	12	8	26
Raft	14	6	8	20	48
Ducks	18	19	15	36	78

ment of them, with the escape route of duplication up your sleeve.

From the point of view of engineering artistry of style, it presents a fascinating concept; a multi-medium torque converter. In a wave the individual drops of water rotate; the ducks collect their energy in angular motion. This is transferred through 'action at a distance' to the precession of a gyroscope of up to 180°, and then transposed into little pumps. We change the grandfather clock into a dentist's drill, and we do it without a gearwheel or even a torque transmitting shaft.

Perhaps it would be interesting to compare the three different wave energy designs development prospects in the form of a table. For instance, we could give a maximum of 20 for the external compatibility of the main invention, and another 20 for the internal. We would use the composite figure of up to 20, as already defined, for the average of the mini-inventions, and 40 for style. This would give us a possible total of 100. My guess is that the figures in Table 7.1 would be reasonable.

We must not forget that the decision of the best design to develop may well be dominated by the forecast of the capital and running costs of the various devices. I can find no published and comparable forecasts for these. Even if they were available they would concern business economics and not research strategy; this book is about travelling hopefully, not forecasting arrivals. We have enough troubles of our own. But, more to the point, I have taken these examples as hypothetical frameworks on which to exhibit the various points I have been making. By the time this book is published the designs may all be improved and the implied criticisms out of date, but this is irrelevant to our present purpose.

8

To be or not to be

So far we have been discussing how to go on with research; but we must say something about not going on with it. As we have seen, many successful inventions have been re-inventions; the first attempt at development failed. In my view there are three main causes for such unhappy occurrences.

The first is that you run out of money. You don't decide to stop the development, your bank manager (or his equivalent) does it for you. Perhaps you had forgotten, or never realised, that development costs are always between three and nine times greater than one's initial conservative estimates.

The second reason for failure is that you have too much money. I can hear squeaks of disbelief tinged with envy ringing in my ears. But it is true. True when people imagine that money is a substitute for thought. And they do. Quite often.

In the last war, money was no problem for secret development work. By and large it didn't make it any easier to do; at any rate for me to do. And today big corporations sometimes think that they can flatten awkward development hurdles by mere money, and when they see they can't, they snatch all the money away again in a huff. I must repeat, there is no substitute for thought.

Talking over the development of new ideas with someone with an international reputation for being successful at it, and who could simultaneously handle a number of different projects, I asked him how he did it. One of the things he told me was this. When research had to be done in a hurry (and when does it not have to be?), he never used to fix a maximum time for his staff to come up with an answer. He used to set a minimum. He would say, for example, 'Go away and do your research and I don't want to hear from you for at least two months'. He knew it would need careful thought, and you can't think that way with one eye on the clock. He demonstrated that although money was no substitute for thought, thought was a good substitute for money.

But, and in this lies the third possibility of failure, thought alone is not enough; it must be patterned thought. A good innovator is not automatically a good research worker. In my experience he is more likely to be automatically a bad one, although I have met some brilliant exceptions.

The best safeguard is to have a map laid out in advance of the various routes you can take. This book is mine. You may well have your own and a better one. The fatal thing is to have none.

EPILOGUE

So far this book has, perhaps, been suffocatingly materialistic. Money in and money out has been its theme. But most of the great inventors of history have not been conditioned in this way; self-interest has not been their driving ambition. They have sought to help the community they live in, and the future they may never see, even though it has meant much personal self-sacrifice.

Some time ago I was sitting in the garden of Emmanuel College, Cambridge, on a summer evening looking again at the flowers, the pond and the ducks, that I had watched from the window of my rooms as a student. The flowers and the ducks were new, but little else had changed. Presently a very old man came wandering in and sat down beside me. Introducing himself, he said 'My name is Jones – Melville Jones'. I told him my name and made small talk while I wildly searched my memory to connect up his name. Suddenly it dawned; my companion was Professor Sir Melville Jones FRS. I told him that he had taught me aeronautics a long time ago. I said 'I still remember much of what you said', and then, after a pause, I added 'especially one sentence you said in a lecture. I apologise if I am being both silly and impertinent, but for some reason I can recall perfectly the way you said it. You were talking about the pioneer days of flying and what you said was "Some of the finest pilots that ever lived killed themselves trying to find a cure for spin in aircraft!" The recollection of your voice saying that has haunted the back of my memory for nearly thirty years. Is there a reason for it or am I just imagining things?'

He said 'I think there is a reason for it', and he told me why. In 1910, he and his brother were greatly concerned about the high risk of flying, and in particular the uncontrollable spinning nose-dive which caused so many deaths. They decided to dedicate their lives to finding the cause and cure. They began their experiments, and when war broke out in 1914 they both joined the RAF and continued their experiments as they flew and fought. After the war ended they were given planes and an airfield to continue their research. One day his brother was on an experimental flight, with the aircraft spinning down towards the ground. He watched and waited for the plane to level out. It never did. It buried itself in the ground a few yards from where he stood, killing his brother instantly. And then he told me 'As I looked down on what remained of my brother I said to myself "I will beat this thing".' And then after a pause, he turned to me and added 'And I did'.

And I am not likely to forget that either. He was, perhaps more than anyone, responsible for making flying safe – a reward infinitely greater than the honours that were heaped upon him.

INDEX